T0192113

Communications in Computer and Information Science 659

Commenced Publication in 2007
Founding and Former Series Editors:
Alfredo Cuzzocrea, Dominik Ślęzak, and Xiaokang Yang

More information about this series at http://www.springer.com/series/7899

Tadeusz Czachórski · Erol Gelenbe
Krzysztof Grochla · Ricardo Lent (Eds.)

Computer and Information Sciences

31st International Symposium, ISCIS 2016
Kraków, Poland, October 27–28, 2016
Proceedings

Springer

Editors
Tadeusz Czachórski
Institute of Theoretical and Applied
  Informatics, Polish Academy of Sciences
Gliwice
Poland

Krzysztof Grochla
Institute of Theoretical and Applied
  Informatics, Polish Academy of Sciences
Gliwice
Poland

Erol Gelenbe
Department of Electrical and Electronic
  Engineering
Imperial College
London
UK

Ricardo Lent
University of Houston
Houston, TX
USA

ISSN 1865-0929   ISSN 1865-0937 (electronic)
Communications in Computer and Information Science
ISBN 978-3-319-47216-4   ISBN 978-3-319-47217-1   (eBook)
DOI 10.1007/978-3-319-47217-1

Library of Congress Control Number: 2016935965

This Springer imprint is published by Springer Nature
The registered company is Springer International Publishing AG
The registered company address is: Gewerbestrasse 11, 6330 Cham, Switzerland

Preface

The 31st International Symposium on Computer and Information Sciences was held during October 27–28, 2016, in Kraków, Poland, under the auspices of the Institute of Theoretical and Applied Informatics of the Polish Academy of Sciences, Gliwice and of Imperial College, London.

This was the 31st event in the ISCIS series of conferences that have brought together computer scientists from around the world, including Ankara, Izmir, and Antalya in Turkey, Orlando, Florida, Paris, London, and Kraków. Thus this conference follows the tradition of very successful previous annual editions, and most recently ISCIS 2015, ISCIS 2014, ISCIS 2013, ISCIS 2012, ISCIS 2011, and ISCIS 2010. The proceedings of previous editions have been included in major research indexes, such as ISI WoS, DBLP, and Google Scholar.

ISCIS 2016 included three invited keynote presentations by leading contributors to the field of computer science, as well as peer-reviewed contributed research papers. The program was established from the submitted papers, and covered relevant and timely aspects of computer science and engineering research, with a clear contribution presenting experimental evidence or theoretical developments and proofs that support the claims of the paper.

The topics included in this year's edition included computer architectures and digital systems, algorithms, theory, software engineering, data engineering, computational intelligence, system security, computer systems and networks, performance modelling and analysis, distributed and parallel systems, bioinformatics, computer vision, and significant applications such as medical informatics and imaging. All the accepted papers were peer reviewed by two or three referees and evaluated on the basis of technical quality, relevance, significance, and clarity.

The organizers and proceedings editors thank the dedicated Program Committee members and other reviewers for their contributions, and would especially like to thank all those who submitted papers, even though only a fraction could be accepted. We also thank Springer for producing these high-quality proceedings of ISCIS 2016.

September 2016

Tadeusz Czachorski
Erol Gelenbe
Krzysztof Grochla
Ricardo Lent

Organization

The 31st International Symposium on Computer and Information Sciences (ISCIS 2016) was organized by the Institute of the Theoretical and Applied Informatics of Polish Academy of Sciences, Gliwice, and Imperial College, London.

Conference Chair

Erol Gelenbe   Imperial College, UK

Program Committee Co-chairs

Lale Akarun   Bogazici University, Turkey
Mehmet Baray   Bilkent University, Turkey
Tadeusz Czachórski   IITiS PAN, Poland
Attila Gursoy   Koç University, Turkey
Albert Levi   Sabanci University, Turkey
Sema Oktug   ITU
Adnan Yazici   METU

Organizing Committee

Krzysztof Grochla   IITiS PAN, Poland
  (Chair)
Konrad Połys   IITiS PAN, Poland
Mariusz Słabicki   IITiS PAN, Poland
Michał Gorawski   IITiS PAN, Poland
Sławomir Nowak   IITiS PAN, Poland

Program Committee

Ethem Alpaydın   Bogaziçi University,Turkey
Cevdet Aykanat   Bilkent University, Turkey
Manfred Broy   TUM
Fazli Can   Bilkent University, Ankara, Turkey
Sophie Chabridon   Institut Telecom, Telecom Sud Paris, France
Tadeusz Czachorski   IITiS of Polish Academy of Science, Poland
Gökhan Dalkılıç   Dokuz Eylul University, Turkey
Mariangiola Dezani   Università di Torino, Italy
Nadia Erdogan   Istanbul Technical University, Turkey
Taner Eskil   Isik University, Turkey
Jean-Michel Fourneau   University of Versailles, France
Stephen Gilmore   University of Edinburgh, UK

Krzysztof Grochla	Institute of Theoretical and Applied Informatics of PAS, Poland
Adam Grzech	Wroclaw University of Technology, Poland
Ugur Güdükbay	Bilkent University, Turkey
Attila Gursoy	Koç University, Turkey
Yorgo Istefanopulos	Isik University, Turkey
Alain Jean-Marie	LIRMM University of Montpellier, France
Sylwester Kaczmarek	Gdansk University of Technology, Poland
Jacek Kitowski	AGH University of Science and Technology, Poland
İbrahim Körpeoğlu	Bilkent University, Turkey
Stefanos Kollias	NTUA Athens, Greece
Jerzy Konorski	Gdansk University of Technology, Poland
Ricardo Lent	University of Houston, USA
Albert Levi	Sabanci University, Turkey
Peixiang Liu	Nova Southeastern University, USA
Jozef Lubacz	Warsaw University of Technology, Poland
Chris Mitchell	Royal Holloway, University of London, UK
Marek Natkaniec	AGH University of Science and Technology, Poland
Sema Oktug	Istanbul Technical University, Turkey
Ender Özcan	University of Nottingham, UK
Oznur Ozkasap	Koc University, Turkey
Ferhan Pekergin	Université Paris 13 Nord, France
Nihal Pekergin	LACL, Université Paris-Est Val de Marne, France
Yves Robert	ENS Lyon, France
Alexane Romariz	Universidade de Brasilia, Brazil
Georgia Sakellari	Greenwich University, UK
Aneas Stafylopatis	National Technical University of Athens, Greece
Halina Tarasiuk	Technical University of Warsaw, Poland
Nigel Thomas	University of Newcastle upon Tyne, UK
Hakki Toroslu	Middle East Technical University, Turkey
Dimitrios Tzovaras	Informatics and Telematics Institute/Centre for Research and Technology Hellas, Greece
Ozgur Ulusoy	Bilkent University, Turkey
Krzysztof Walkowiak	Wroclaw University of Technology, Poland
Wei Wei	Xi'an University of Technology, China
Jozef Wozniak	Gdansk University of Technology, Poland
Zhiguang Xu	Valdosta State University, USA
Emine Yilmaz	Microsoft Research, Cambridge, UK
Qi Zhu	University of Houston-Victoria, USA
Thomas Zeugmann	Hokkaido University, Japan

Additional Reviewers

Giuliana Franceschini
Tugrul Dayar
Thanos Thanos
Georgios Stratogiannis

Contents

Smart Algorithms

Data Classification and Processing

Stochastic Modelling

Performance Evaluation

Queuing Systems

Wireless Networks and Security

Image Processing and Computer Vision

Smart Algorithms

An Adaptive Heuristic Approach for the Multiple Depot Automated Transit Network Problem

Olfa Chebbi[(✉)], Ezzeddine Fatnassi, and Hadhami Kaabi

Institut Supérieur de Gestion de Tunis, Université de Tunis,
41, Rue de la Liberté-Bouchoucha, 2000 Bardo, Tunisia
olfaa.chebbi@gmail.com

Abstract. Automated Transit Networks (ATN) are innovative transportation systems where fully driverless vehicles offer an exclusive ondemand transportation service. Within this context of ATN, this study tries to deal with a specific routing problem arising in the context of a ATN's network with a multiple depot topology. More specifically, we present an optimization routing model for automated transit networks which can be used to strategically evaluate depots locations. Our model extends the basic Multi-depot Vehicle Routing Problem (MDVRP). In this paper, the proposed model is tackled using an heuristic approach as the proposed problem is NP-Hard. Experiments are run on a carefully generated instances based on the works from the literature. The numerical results show that the proposed algorithm is competitive as it founds a small gap relative to a lower bound values from the literature.

Keywords: Automated transit network · Multi-depot vehicle routing problem · Heuristics · Genetic algorithm

1 Introduction

Nowadays, public rapid transit systems provide an interesting way for reducing the distinctive negative impact of transportation tools in urban areas. In fact, public rapid transit systems help to improve the access of lower income groups in societies to transportation tools as well as reducing the environmental impact of urban mobilities. Public rapid transit systems consists of light rapid transit (LRT), bus rapid transit (BRT), Automated Transit Networks (ATN), metro, commuters rail and so on. Recently, several models has been put forward to justify the operational, tactical and strategic implementation of rapid transit systems. In this paper, we focus on the implementation of ATN. We extend the operational model of Mrad and Hidri [10] which is used as a base of our operational ATN model.

In the operational model of Mrad and Hidri [10], the optimized variables are the energy consumption, the objective function is the minimization of total

© The Author(s) 2016
T. Czachórski et al. (Eds.): ISCIS 2016, CCIS 659, pp. 3–11, 2016.
DOI: 10.1007/978-3-319-47217-1_1

energy consumption of ATN. The ATN'network is assumed to have a single un-capacitated depot [7]. We extend this model to account for a multiple depot topology network. We introduce also a maximum allowable distance constraint related to the electric battery capacity of the ATN vehicles. We study the effect of these constraints on the operational level for the multiple depot topology ATN'network. In spite of its relative complexity, the proposed operational model could be solved heuristically based on approximate methods which could yields some analytical insight on the structure of its optimal solutions. In particular, we found that introducing multiple depots topology helps to reduce the total service time for rapid transit users. Also, the proposed heuristic approach was proven to found good quality solutions in a fast computational time.

The remainder of this paper is as follows: Sect. 2 presents the ATN system and its related literature review which motivates our work. Section 3 presents the optimization model. Our proposed heuristic approach is introduced in Sect. 4. Section 5 provides numerical results analysis of our approach. Conclusions are reported in Sect. 6.

2   The Automated Transit Networks

ATN (also called Personal Rapid Transit (PRT)) consists mainly on a set of small automated driverless electrical vehicles running on a set of exclusive guideways. ATN is implemented to provide an interesting mode of urban transportation service which could address the need of urban mobility based on specific set-tings. Table 1 provides an overview of the several needs related to urban mobility and how could ATN satisfy them. In the literature, there is a general consen-sus that the key characteristics of ATN includes [2]: (i) Fully automated vehi-cles; (ii) Small and dedicated guideways; (iii) On-demand, origin-to-destination service; (iv) Off-line stations; and (v) A network or system of fully connected guideways.

Table 1. ATN main features

Need	ATN feature
Provide faster service	Non-stop, on-demand service
Reduce congestion	Faster and personalized service to attract private automobile users
Reduce pollution	Electric vehicles
Reduce energy use	Small vehicles
On-demand and Non-stop transportation service to eliminate empty vehicle movements	

2.1   Literature Review

ATN as a conceptual mode of public rapid transit systems has a history of over 60 years. Since, its first introduction in 1953 [2], it was studied by governments, universities, research organization and so on. Literature of ATN includes several books, scholar papers and technical reports. These studies proposed to treat several features related to ATN such as technical and operational analysis, system design, environmental impact, cost performance and so on.

A literature review published in 2005 [5] states that there is more than 200 research papers related to ATN. More recently, several operational and strategic optimization studies related to ATN were published such as simulation [7], energy minimization [10], total traveled distance [6,8], optimized operational planning [4] and so on. However and from our literature review, many optimization routing models related to ATN considered a single depot network topology [4,6,10].

Consequently, it becomes of a high interest to study optimization routing problem related to ATN based on a multiple network topology. Therefore in the next section, we extends the single depot based optimization model of Mrad and Hidri [10] to propose a multiple depot optimization model which would aim at reducing the total travel time of ATN vehicles while serving a set of known static deterministic list of passengers travels.

3   The Optimization Model

In this section, we present the multiple depots ATN optimization model which extends the works of Mrad and Hidri [10]. We first start by presenting the set of assumptions related to our model. Then, we give a graph based model. Finally, we present the complexity of our problem.

3.1   The Set of Assumptions

Let suppose that we have a ATN N with a finite number of stations M. N satisfies connectivity constraints. Therefore, a ATN vehicle could travel between any pairs of stations in N. We suppose that N has a set of depots $\kappa = \{d_1, d_2, d_3, ...d_k\}$ where k represents the number of depot in N. In each depot, there exists an unlimited number of ATN vehicles. The exact number of vehicles needed from each depot is considered as a decision variable. Each vehicle has a limited battery capacity denoted B. We supposed to have a static pre-deterministic list of trips to serve denoted T. $|T| = n$. Each trip i is identified by a quadruplet:

   (i)   a depart time Dt_i,
  (ii)   a depart station Ds_i
 (iii)   an arrival time At_i and
  (iv)   an arrival station As_i

Finally, let SP be a matrix cost which defines the shortest time travel path between each pair of stations.

3.2 Graph Based Formulation

Our problem has an objective to find a set of least cost roads starting and ending at one of the depots in N which minimizes the total travel time of the ATN vehicle while serving each trip exactly once. To model our problem, let us define $G = \{V, E\}$ where V is a set of nodes and E is a set of edges. Each trip i is represented by a node in V. Also, each depot d_i is represented by two nodes s_i and t_i. Also, we have n trips and k depots. The cardinality of V is equal to $n + 2k$. $V^* = V \backslash \{s_1, s_2, ... s_k, t_1, t_2, ..., t_k\}$. As for the set of edge E, it will be defined as following:

- For each pair of nodes i and $j \in V^*$, we add an edge (i, j) to E if $At_i + SP_{(As_i, Ds_j)} \leq Dt_j$. The edge has a cost c_{ij}, representing the total time needed to move from arrival station As_i of trip i to depart station Ds_j of trip j.
- For each node i and each depot k, we add an edge (k, i). This edge has a cost c_{ki} which is equal to total traveled time to reach the depart station Ds_i of trip i, from the depot k.
- For each node i and each depot k, we add an edge (i, k). This edge has as a cost the total travel time needed to move from the arrival station As_i of trip i to the depot k.

Let us also denote $E^* = E\{(i, j)$ where $i \in \kappa$ or $j \in \kappa\}$.

3.3 The Complexity of Our Problem

Starting from our graph modeling of the problem, we could note that it extends the asymmetric distance constrained vehicle routing problem (ADCVRP) [1]. Our problem is asymmetric as the cost of edge $(i, j) \neq (j, i)$. The DCVRP is a vehicle routing problem where each road is subject to total distance, time or cost constraints. The ADCVRP is not well studied in the literature. In fact and as Almoustafa et al. state [1], only two papers studied this problem [1]. The work related to ADCVRP are based on a single depot topology. Therefore, our proposed ATN problem could be considered as an extension to the ADCVRP by adding multiple depots to its basic version. Thus, it represents an interesting worth to study extension to the works in the literature. The ADCVRP is proven to be an NP-Hard problem [8]. Consequently, our proposed extension to the ADCVRP is an NP-Hard problem. In the next section, we present details of our solution approach proposed to solve our problem.

4   Genetic Algorithm Approach

As mentioned earlier, the proposed multiple depot ATN routing problem is an NP-Hard optimization problem which has its own difficulties to solve. Consequently, this paper presents an heuristic approach based on the implementation of genetic algorithm (GA) to solve the proposed optimization problem. GA presents a good solution approach for the proposed ATN problem as it could discover many different zones in the search space [4]. Consequently, it could reach

---

Algorithm 1. Pseudo Code of Genetic Algorithm

1: Initialize-parameter()
2: **while** Not reach termination criterion **do**
3:    **for all** Individual in the population **do**
4:       parent1 ⟵ Select-at-random(pop)
5:       parent2 ⟵ Select-at-random(pop)
6:       offspring ⟵ One point Crossover(parent1, parent2)
7:       offspring ⟵ Insertion mutation(offspring)
8:       Evaluate(offspring)
9:       **if** the offspring is better than the worst individual **then**
10:          The offspring replace the worst individual in the population)
11:       **end if**
12:    **end for**
13: **end while**
14: individual ⟵ Best-individual(pop)

---

a good quality solution in a fast computational time. A high level overview of our GA is presented in Algorithm 1.

The choice of developing GA[1] for this problem is motivated by the fact that large number of studies adopted this solution approach to solve routing problems. One could note for instance [9].

Similarly and starting from a population of individuals, a GA applies genetic operators like crossover and mutation in each iteration in order to generate new offsprings. Consequently, the key issue to successfully develop GA is to select the appropriate genetic operators and solution representation.

In the next subsections, we focus more closely on the proposed GA. We first describe the individual'representation and evaluation function. Then, we discuss the implemented genetic operators and the parameters used therein.

4.1 Solution Representation and Evaluation Function

In our GA, a solution is represented using a vector of trips to perform. In this vector, each trip is represented by a single gene only once. Therefore, each solution is in a form of a permutation of trips. As for the evaluation function, we adapt the split function of Prins [11] to our context. More specifically and starting from a permutation, the split function constructs an auxiliary graph where each node represents a trip in addition to a node representing the different depots in the ATN'network. Each edge in the auxiliary graph represents a feasible road based in the permutation at hand. Next, the algorithm uses the shortest path in the auxiliary graph to find the related set of roads. Thus, we obtain the set of roads starting and ending at one of the depots in the network covering each trip only once. More details could be found in [11].

---

[1] Non expert readers can for instance refer to [12] for more details about GA.

4.2   Crossover and Mutation Operators

After deciding the representation form of the individuals in the GA, two parents are selected randomly according to Algorithm 1 in order to create new offsprings using crossover operator. Our crossover operator applied in our algorithm is the one point crossover. For the first parent, we choose randomly a cut point. The trips that are present before the cut point in the first parent are copied to the offspring. The missing trips in the resulted offspring are copied from the second parent while following their order of appearance. More details could be found in Fig. 1.

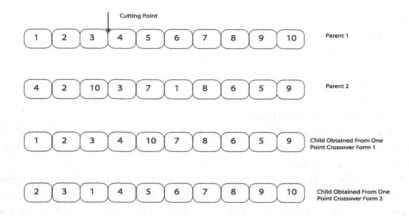

Fig. 1. Example of one point crossover

Also, mutation helps GAs to preserve diversification in the population. In our algorithm, the mutation procedure is applied on the new generated offspring after the crossover operator. In our approach, we use the insertion mutation operator. This operator chooses at random one trip from the permutation and insert it at a random position.

5   Computational Results

In this section, we present the computational results related to the proposed GA. The algorithms proposed in this paper were coded in C++ language. The experiments are performed on a PC with a 3.2 GHZ CPU and 8 GB of RAM.

5.1   Test Instances

To test our proposed approach, we generated 100 ATN multiple depot instances. The size of the problem (i.e. the number of trips) in our testing bed varies between 10 and 100 trips by a step of 10. For each number of trips, 10 instances were generated. To generate the different instances, the ATN's instances generator from the literature of Mrad and Hidri [10] is adapted to our context. To assert the quality of the obtained solutions we used the GAP metric. The GAP is obtained as follows:

$$\text{GAP} = \left(\frac{(SOL - LB)}{LB}\right) \times 100 \tag{1}$$

We should note that SOL is the solution of LB represents the linear relaxation of the valid mathematical formulation presented in the literature [1]. The mathematical models related to the linear relaxation were implemented using the IBM ILOG CPLEX Optimizer 12.2.

5.2 Result of the Genetic Algorithm

As for the parameter tuning, we used a specific method from the literature to effectively tune our proposed GA [3]. Based on this method, we found the following parameters: (i) Number of generations:800, (ii) population size:20; (iii) crossover rate:0.9 and (iv) mutation rate: 0.3. Table 2 presents the results of our approach. It should be noted the good quality of our proposed GA as we found an average GAP of 2.859 % in 0.231 s.

We should note also that the average GAP grows steadily. The maximum GAP was equal to 6.435 % which is still represents good quality results. As for the average time, our algorithm proved to be very effective as the average computational time was still below 1 s. These results comfort our choice in the selection of a GA for solving our hard combinatorial optimization problem related to ATN. These results are encouraging in term of problem solvability.

Table 2. The Obtained Results

Number of travels	Average GAP %	Average time in seconds
10	0	0.833
20	0.575	0.039
30	0.438	0.485
40	0.832	0.063
50	1.771	0.082
60	3.327	0.103
70	4.263	0.122
80	4.628	0.148
90	6.327	0.191
100	6.435	0.241
Average	2.859	0.231

6 Conclusions

In this paper the Multi-Depot automated transit network problem is evoked and modeled. A genetic algorithm is proposed and implemented to solve it. The proposed algorithm integrates an effective genetic operators and evaluation function for solving the combinatorial optimization problem. The algorithm constructs a set of ATN'vehicles routes starting and ending at any of the proposed depots with minimum routing costs. Computational experiments on a set of carefully generated instances show that the proposed heuristic is very effective. As an extension to this work, a more adapted meta-heuristic approach such as bee colony algorithms, ant colony algorithm could be adapted to our context. Also the inclusion of additional constraints such as mixed fleet with varying maximum allowable distance and multi-compartment vehicles is under investigation.

References

1. Almoustafa, S., Hanafi, S., Mladenovi, N.: New exact method for large asymmetric distance-constrained vehicle routing problem. Eur. J. Oper. Res. (2012)
2. Carnegie, J.A., Hoffman, P.S.: Viability of personal rapid transit in New Jersey. Technical report (2007)
3. Chebbi, O., Chaouachi, J.: Effective parameter tuning for genetic algorithm to solve a real world transportation problem. In: 2015 20th International Conference on Methods and Models in Automation and Robotics (MMAR), pp. 370–375. IEEE (2015)
4. Chebbi, O., Chaouachi, J.: Reducing the wasted transportation capacity of personal rapid transit systems: an integrated model and multi-objective optimization approach. Transp. Res. Part E Logistics Transp. Rev. **89**, 236–258 (2015)
5. Cottrell, W.D.: Critical review of the personal rapid transit literature. In: Proceedings of the 10th International Conference on Automated People Movers, pp. 1–4, May 2005
6. Fatnassi, E., Chebbi, O., Chaouachi, J.: Discrete honeybee mating optimization algorithm for the routing of battery-operated automated guidance electric vehicles in personal rapid transit systems. Swarm Evol. Comput. **26**, 35–49 (2015)
7. Fatnassi, E., Chebbi, O., Siala, J.C.: Two strategies for real time empty vehicle redistribution for the personal rapid transit system. In: 2013 16th International IEEE Conference on Intelligent Transportation Systems (ITSC), pp. 1888–1893. IEEE (2013)

8. Fatnassi, E., Chebbi, O., Siala, J.C.: Comparison of two mathematical formulations for the offline routing of personal rapid transit system vehicles. In: The International Conference on Methods and Models in Automation and Robotics (2014)
9. Lahyani, R., Khemakhem, M., Semet, F.: Rich vehicle routing problems: from a taxonomy to a definition. Eur. J. Oper. Res. **241**(1), 1–14 (2015)
10. Mrad, M., Hidri, L.: Optimal consumed electric energy while sequencing vehicle trips in a personal rapid transit transportation system. Comput. Ind. Eng. **79**, 1–9 (2015)
11. Prins, C., Lacomme, P., Prodhon, C.: Order-first split-second methods for vehicle routing problems: a review. Transp. Res. Part C Emerging Technol. **40**, 179–200 (2014)
12. Sivanandam, S., Deepa, S.: Introduction to Genetic Algorithms. Springer Science & Business Media, New York (2007)

An Analysis of the Taguchi Method for Tuning a Memetic Algorithm with Reduced Computational Time Budget

Düriye Betül Gümüş[✉], Ender Özcan, and Jason Atkin

ASAP Research Group, School of Computer Science, University of Nottingham,
Wollaton Road, Nottingham NG8 1BB, UK
{betul.gumus,ender.ozcan,jason.atkin}@nottingham.ac.uk

Abstract. Determining the best initial parameter values for an algorithm, called parameter tuning, is crucial to obtaining better algorithm performance; however, it is often a time-consuming task and needs to be performed under a restricted computational budget. In this study, the results from our previous work on using the Taguchi method to tune the parameters of a memetic algorithm for cross-domain search are further analysed and extended. Although the Taguchi method reduces the time spent finding a good parameter value combination by running a smaller size of experiments on the training instances from different domains as opposed to evaluating all combinations, the time budget is still larger than desired. This work investigates the degree to which it is possible to predict the same good parameter setting faster by using a reduced time budget. The results in this paper show that it was possible to predict good combinations of parameter settings with a much reduced time budget. The good final parameter values are predicted for three of the parameters, while for the fourth parameter there is no clear best value, so one of three similarly performing values is identified at each time instant.

Keywords: Evolutionary algorithm · Parameter tuning · Design of experiments · Hyper-heuristic · Optimisation

1 Introduction

Many real-world optimisation problems are too large for their search spaces to be exhaustively explored. In this research we consider cross-domain search where the problem structure will not necessarily be known in advance, thus cannot be leveraged to produce fast exact solution methods. Heuristic approaches provide potential solutions for such complex problems, intending to find near optimal solutions in a significantly reduced amount of time. Metaheuristics are problem-independent methodologies that provide a set of guidelines for heuristic optimization algorithms [18]. Among these, memetic algorithms are highly effective population-based metaheuristics which have been successfully applied to

© The Author(s) 2016
T. Czachórski et al. (Eds.): ISCIS 2016, CCIS 659, pp. 12–20, 2016.
DOI: 10.1007/978-3-319-47217-1_2

a range of combinatorial optimisation problems [2,8,10,11,14]. Memetic algorithms, introduced by Moscato [12], hybridise genetic algorithms with local search. Recent developments in memetic computing, which broadens the concept of memes, can be found in [13]. Both the algorithm components and parameter values need to be specified in advance [17], however determining the appropriate components and initial parameter settings (i.e., parameter tuning) to obtain high quality solutions can take a large computational time.

Hyper-heuristics are high-level methodologies which operate on the search space of low-level heuristics rather than directly upon solutions [4], allowing a degree of domain independence where needed. This study uses the Hyper-heuristics Flexible Framework (HyFlex) [15] which provides a means to implement general purpose search methods, including meta/hyper-heuristics.

In our previous work [7], the parameters of a memetic algorithm were tuned via the Taguchi method, under a restricted computational budget, using a limited number of instances from several problem domains. The best parameter setting obtained through the tuning process was observed to generalise well to unseen instances. A drawback of the previous study was that even testing only the 25 parameter combinations indicated by the L_{25} Taguchi orthogonal array, still takes a long time. In this study, we further analyse and extend our previous work with an aim to assess whether we can generalise the best setting sooner with a reduced computational time budget. In Sect. 2, the HyFlex framework is described. Our methodology is discussed in Sect. 3. The experimental results and analysis are presented in Sect. 4. Finally, some concluding remarks and our potential future work are given in Sect. 5.

2   Hyper-Heuristics Flexible Framework (HyFlex)

Hyper-heuristics Flexible Framework (HyFlex) is an interface proposed for the rapid development, testing and comparison of meta/hyper-heuristics across different combinatorial optimisation problems [15]. There is a logical barrier in HyFlex between the high-level method and the problem domain layers, which prevents hyper-heuristics from accessing problem specific information [5]. Only problem independent information, such as the objective function value of a solution, can pass to the high-level method [3].

HyFlex was used in the first Cross-domain Heuristic Search Challenge (CHeSC2011) for the implementation of the competing hyper-heuristics. Twenty selection hyper-heuristics competed at CHeSC2011. Details about the competition, the competing hyper-heuristics and the tools used can be found at the CHeSC website[1]. The performance comparison of some previously proposed selection hyper-heuristics including one of the best performing ones can be found in [9]. Six problem domains were implemented in the initial version of HyFlex: Maximum Satisfiability (MAX-SAT), One Dimensional Bin Packing (BP), Permutation Flow Shop (PFS), Personnel Scheduling (PS), Traveling Salesman (TSP) and Vehicle Routing (VRP). Three additional problem domains were

---

[1] http://www.asap.cs.nott.ac.uk/external/chesc2011/.

added by Adriaensen et al. [1] after the competition: 0-1 Knapsack (0-1 KP), Max-Cut, and Quadratic Assignment (QAP). Each domain contains a number of instances and problem specific components, including low level heuristics and an initialisation routine which can be used to produce an initial solution. In general, this routine creates a random solution.

The low-level heuristics (operators) in HyFlex are categorised as *mutation*, *ruin and re-create*, *crossover* and *local search* [15]. Mutation makes small random perturbations to the input solution. Ruin and re-create heuristics remove parts from a complete solution and then rebuild it, and are also considered as mutational operators in this study. A crossover operator is a binary operator accepting two solutions as input unlike the other low level heuristics. Although there are many crossover operators which create two new solutions (*offspring*) in the scientific literature, the Hyflex crossover operators always return a single solution (by picking the best solution in cases where the operator produces two offspring). Local search (hill climbing) heuristics iteratively perform a search within a certain neighbourhood attempting to find an improved solution. Both local search and mutational heuristics come with parameters. The *intensity of mutation* parameter determines the extent of changes that the mutation or ruin and re-create operators will make to the input solution. The *depth of search* parameter controls the number of steps that the local search heuristic will complete. Both parameter values vary in [0,1]. More details on the domain implementations, including low level heuristics and initialisation routines can be found on the competition website and in [1,15].

3   Methodology

Genetic algorithm are well-known metaheuristics which perform search using the ideas based on natural selection and survival of the fittest [6]. In this study, a steady state memetic algorithm (SSMA), hybridising genetic algorithms with local search is applied to a range of problems supported by HyFlex, utilising the provided mutation, crossover and local search operators for each domain.

SSMA evolves a *population* (set) of initially created and improved *individuals* (candidate solutions) by successively applying genetic operators to them at each evolutionary cycle. In SSMA, a fixed number of individuals, determined by the *population size* parameter, are generated by invoking the HyFlex initialisation routine of the relevant problem domain. All individuals in the population are evaluated using a *fitness function* measuring the quality of a given solution. Each individual is improved by employing a randomly selected local search operator. Then the evolutionary process starts. Firstly, two individuals are chosen one at a time for crossover from the current population. The generic *tournament selection* which chooses the fittest individual (with the best fitness value with respect to the fitness function) among a set of randomly selected individuals of *tournament size* (*tour size*) is used for this purpose. A randomly chosen crossover operator is then applied producing a single solution which is perturbed using a randomly selected mutation and then improved using a randomly selected local search.

Finally, the resultant solution gets evaluated and replaces the worst individual in the current population. This evolutionary process continues until the time limit is exceeded.

SSMA has parameters which require initial settings and influence its performance. Hence, the Taguchi orthogonal arrays method [16] is employed here to tune these parameter settings. Firstly, control parameters and their potential values (levels) are determined. Four algorithm parameters are tuned: population size (PopSize), tournament size (TourSize), intensity of mutation (IoM) and depth of search (DoS). The parameter levels of $\{0.2, 0.4, 0.6, 0.8, 1.0\}$ are used for both IoM and DoS. PopSize takes a value in $\{5, 10, 20, 40, 80\}$. Finally, $\{2, 3, 4, 5\}$ are used for TourSize. HyFlex ensures that these are problem independent parameters, i.e. common across all of the problem domains. Based on the number of parameters and levels, a suitable orthogonal array is selected to create a design table. Experiments are conducted based on the design table using a number of 'training' instances from selected domains and then the results are analysed to determine the optimum level for each individual control parameter. The combination of the best values of each parameter is predicted to be the best overall setting.

4   Experimentation and Results

In [7], experiments were performed with a number of configurations for SSMA using 2 training instances from 4 HyFlex problem domains. An execution time of 415 seconds was used as a termination criterion for those experiments, equivalent to 10 nominal minutes on the CHeSC2011 computer, as determined by the evaluation program provided by the competition organisers. Each configuration was tested 31 times, the median values were compared and the top 8 algorithms were assigned scores using the (2003–2009) Formula 1 scoring system, awarding 10, 8, 6, 5, 4, 3, 2 and 1 point(s) for the best to the 8th best, respectively. The best configuration was predicted to be $IoM = 0.2$, $DoS = 1.0$, $TourSize = 5$ and $PopSize = 5$, and this was then applied to unseen instances from 9 domains and found to perform well for those as well. A similar process was then applied to predict a good parameter configuration across 5 instances from each of the 9 extended HyFlex problem domains, and the same parameter combination was found, indicating some degree of cross-domain value to the parameter setting. With 31 repetitions of 25 configurations, this was a time-consuming process.

The aim of this study is to investigate whether a less time consuming analysis could yield similar information. All 25 parameter settings indicated by the L_{25} Taguchi orthogonal array were executed with different time budgets, from 1 to 10 min of nominal time (matching the CHeSC2011 termination criterion), the Taguchi method was used to predict the best parameter configuration for each duration and the results were analysed. 2 arbitrarily chosen instances from each of the 6 original HyFlex problem domains were employed during the first parameter tuning experiments. Figure 1 shows the main effect values for each parameter level, defined as the mean total Formula 1 score across all of the

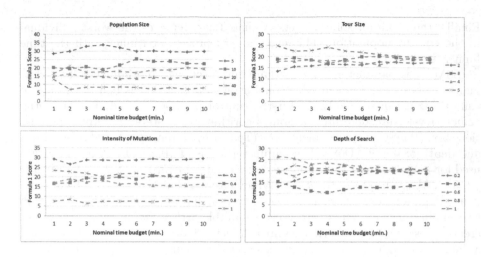

Fig. 1. Main effects of parameter values at different times using 2 training instances from 6 problem domains

settings where the parameter took that specific value. It can be seen that a population size of 5 has the highest effect in each case during the 10 nominal minutes run time. Similarly, the intensity of mutation parameter value of 0.2 performs well at each time. For the tour size parameter, 5 has the highest effect throughout the search except at one point: at 10 nominal minutes, the tour size of 4 had a score of 19.58 while tour size 5 had a score of 19.48, giving very similar results. The best value for the depth of search parameter changes during the execution; however, it is always one of the values 0.6, 0.8 or 1.0. 0.6 for depth of search is predicted to be the best parameter value for a shorter run time.

The analysis of variance (ANOVA) is commonly applied to the results in the Taguchi method to determine the percentage contribution of each factor [16]. This analysis helps the decision makers to identify which of the factors need more control. Table 1 shows the percentage contribution of each factor. It can be seen that intensity of mutation and population size parameters have

Table 1. The percentage contribution of each parameter obtained from the Anova test for 6 problem domains

par. \n.t.b. (min.)	1	2	3	4	5	6	7	8	9	10
IoM	37.6%	22.6%	28.8%	24.6%	28.2%	29.9%	32.4%	32.6%	34.1%	36.3%
DoS	14.8%	13.2%	9.3%	11.0%	9.5%	6.6%	6.3%	6.4%	5.4%	4.0%
PopSize	20.5%	34.0%	35.6%	38.2%	38.5%	38.3%	37.7%	39.4%	39.4%	35.1%
TourSize	10.7%	3.7%	3.2%	5.0%	2.8%	3.0%	2.0%	0.8%	0.8%	0.5%
Residual	16.3%	26.5%	23.0%	21.1%	21.0%	22.2%	21.5%	20.8%	20.2%	24.1%

Table 2. The p-values of each parameter obtained from the Anova test for 6 domains. The parameters which contribute significantly are marked in bold.

par. \n.t.b. (min.)	1	2	3	4	5	6	7	8	9	10
IoM	**0.019**	0.191	0.090	0.105	0.078	0.077	0.060	0.054	**0.045**	0.060
DoS	0.171	0.406	0.497	0.384	0.450	0.633	0.635	0.614	0.669	0.825
PopSize	0.090	0.086	0.056	**0.037**	**0.036**	**0.042**	**0.041**	**0.033**	**0.031**	0.065
TourSize	0.188	0.746	0.741	0.568	0.757	0.749	0.836	0.945	0.947	0.977

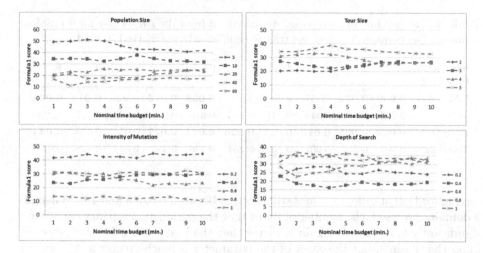

Fig. 2. Main effects of parameter values at different time using 2 training instances from 9 problem domains

highest percentage contribution to the scores. P-values lower than 0.05 means that the parameter is found to contribute significantly to the performance with a confidence level of 95 %. Table 2 shows the p-values of the parameters at each time. The contribution of the PopSize parameter is found to be significant in 6 out of 10 time periods, whereas the intensity of mutation parameter contributes significantly in only 2 out of 10 time periods and the contribution of the other parameters was not found to be significant.

In order to investigate the effect of Depth of Search (DoS) further, we increased the number of domains considered to 9 (and thus used 18 training instances). The main effects of the parameter values are shown in Fig. 2 and Tables 3 and 4 show the percentage contributions and p-values for each parameter. It can be observed from Fig. 2 that the best parameter value does not change over time for the PopSize, TourSize and IoM parameters. The best parameter setting could be predicted for these three parameters after only 1 nominal minute of run time. However, for the depth of search parameter, the best setting indicated in [7] is found only when the entire run time has been used. The best

Table 3. The percentage contribution of each parameter obtained from the Anova test for 9 domains

par. \n.t.b. (min.)	1	2	3	4	5	6	7	8	9	10
IoM	27.7 %	23.6 %	24.0 %	20.3 %	26.3 %	30.0 %	39.1 %	37.3 %	43.4 %	46.0 %
DoS	7.1 %	12.3 %	9.6 %	11.7 %	12.4 %	10.4 %	10.1 %	12.3 %	12.5 %	10.8 %
PopSize	47.3 %	44.5 %	40.8 %	38.2 %	35.3 %	35.6 %	30.9 %	28.3 %	25.0 %	25.5 %
TourSize	8.5 %	7.3 %	9.9 %	14.0 %	8.9 %	7.2 %	4.8 %	4.1 %	3.2 %	2.6 %
Residual	9.4 %	12.3 %	15.7 %	15.8 %	17.1 %	16.7 %	15.1 %	18.1 %	15.9 %	15 %

Table 4. The p-values of each parameter obtained from the Anova test for 9 problem domains. The parameters which contribute significantly are marked in bold.

par. \n.t.b. (min.)	1	2	3	4	5	6	7	8	9	10
IoM	**0.009**	**0.032**	0.057	0.086	0.056	**0.038**	**0.013**	**0.026**	**0.011**	**0.008**
DoS	0.232	0.144	0.317	0.241	0.248	0.310	0.278	0.274	0.217	0.251
PopSize	**0.002**	**0.005**	**0.013**	**0.017**	**0.026**	**0.024**	**0.027**	0.054	0.053	**0.044**
TourSize	0.109	0.219	0.201	0.112	0.263	0.336	0.453	0.587	0.628	0.677

setting for DoS at different times still changes between 0.6, 0.8 and 1.0. When all 9 domains are used, the number of times that the parameters settings contribute significantly is increased. Again it seems that the best setting for DoS depends upon the runtime, but the effect of the parameter is much greater at the longer execution times with the addition of the new domains.

These three values combining with the best values of other parameters were then tested separately on all 45 instances from 9 domains, with the aim of finding the best DoS value on all instances. According to the result of experiments, each of these three configurations found the best values for 18 instances (including ties), considering their median performances over 31 runs. This indicates that these three configurations actually perform similarly even though there are small differences overall. Hence, using only one nominal minute and 2 instances from 6 domains was sufficient to obtain the desired information about the best configuration, reducing the time needed for parameter tuning significantly.

5   Conclusion

This study extended and analysed the previous study in [7], applying the Taguchi experimental design method to obtain the best parameter settings with different run-time budgets. We trained the system using 2 instances from 6 and 9 domains separately and tracked the effects of each parameter level over time. The experimental results show that good values for three of the parameters are relatively easy to predict, but the performance is less sensitive to the value of the fourth (DoS), with different values doing well for different instances and very similar, "good", overall performances for three settings, making it hard to identify a single "good" value. In summary, these results show that it was possible to predict a good parameter combination by using a much reduced time budget for cross domain search.

References

1. Adriaensen, S., Ochoa, G., Nowé, A.: A benchmark set extension and comparative study for the hyflex framework. In: IEEE Congress on Evolutionary Computation, CEC 2015, 25–28 May 2015, Sendai, Japan, pp. 784–791 (2015)
2. Alkan, A., Özcan, E.: Memetic algorithms for timetabling. In: The 2003 Congress on Evolutionary Computation, CEC 2003, vol. 3, pp. 1796–1802. IEEE (2003)
3. Burke, E.K., Gendreau, M., Hyde, M., Kendall, G., McCollum, B., Ochoa, G., Parkes, A.J., Petrovic, S.: The cross-domain heuristic search challenge – an international research competition. In: Coello, C.A.C. (ed.) LION 2011. LNCS, vol. 6683, pp. 631–634. Springer, Heidelberg (2011)
4. Burke, E.K., Gendreau, M., Hyde, M., Kendall, G., Ochoa, G., Özcan, E., Qu, R.: Hyper-heuristics: a survey of the state of the art. J. Oper. Res. Soc. **64**(12), 1695–1724 (2013)
5. Cowling, P.I., Kendall, G., Soubeiga, E.: A hyperheuristic approach to scheduling a sales summit. In: Burke, E., Erben, W. (eds.) PATAT 2000. LNCS, vol. 2079, p. 176. Springer, Heidelberg (2001)
6. Gendreau, M., Potvin, J.Y.: Metaheuristics in combinatorial optimization. Ann. Oper. Res. **140**(1), 189–213 (2005)
7. Gümüş, D.B., Özcan, E., Atkin, J.: An investigation of tuning a memetic algorithm for cross-domain search. In: 2016 IEEE Congress on Evolutionary Computation (CEC). IEEE (2016)
8. Ishibuchi, H., Kaige, S.: Implementation of simple multiobjective memetic algorithms and its applications to knapsack problems. Int. J. Hybrid Intell. Syst. **1**(1), 22–35 (2004)
9. Kheiri, A., Özcan, E.: An iterated multi-stage selection hyper-heuristic. Eur. J. Oper. Res. **250**(1), 77–90 (2015)
10. Krasnogor, N., Smith, J., et al.: A memetic algorithm with self-adaptive local search: TsP as a case study. In: GECCO, pp. 987–994 (2000)
11. Merz, P., Freisleben, B.: A comparison of memetic algorithms, tabu search, and ant colonies for the quadratic assignment problem. In: Proceedings of the 1999 Congress on Evolutionary Computation, CEC 1999, vol. 3, pp. 2063–2070 (1999)

12. Moscato, P., et al.: On evolution, search, optimization, genetic algorithms and martial arts: towards memetic algorithms. Caltech Concur. Comput. Prog. C3P Rep. **826**, 1989 (1989)
13. Neri, F., Cotta, C.: Memetic algorithms and memetic computing optimization: a literature review. Swarm Evol. Comput. **2**, 1–14 (2012)
14. Ngueveu, S.U., Prins, C., Calvo, R.W.: An effective memetic algorithm for the cumulative capacitated vehicle routing problem. Comput. Oper. Res. **37**(11), 1877–1885 (2010)
15. Ochoa, G., et al.: HyFlex: a benchmark framework for cross-domain heuristic search. In: Hao, J.-K., Middendorf, M. (eds.) EvoCOP 2012. LNCS, vol. 7245, pp. 136–147. Springer, Heidelberg (2012)
16. Roy, R.: A Primer on the Taguchi Method. Competitive Manufacturing Series. Van Nostrand Reinhold, New York (1990)
17. Segura, C., Segredo, E., León, C.: Analysing the robustness of multiobjectivisation approaches applied to large scale optimisation problems. In: Tantar, E., Tantar, E.-E., Bouvry, P., Del Moral, P., Legrand, P., Coello-Coello, C.A., Schütze, O. (eds.) EVOLVE-A Bridge between Probability, Set Oriented Numerics and Evolutionary Computation. SCI, vol. 447, pp. 365–391. Springer, Heidelberg (2013)
18. Sörensen, K., Glover, F.W.: Metaheuristics. In: Gass, S.I., Fu, M.C. (eds.) Encyclopedia of Operations Research and Management Science, pp. 960–970. Springer, New York (2013)

Ensemble Move Acceptance in Selection Hyper-heuristics

Ahmed Kheiri[1]([✉]), Mustafa Mısır[2], and Ender Özcan[3]

[1] Operational Research Group, School of Mathematics, Cardiff University,
Senghennydd Road, Cardiff CF24 4AG, UK
KheiriA@cardiff.ac.uk
[2] Nanjing University of Aeronautics and Astronautics,
College of Computer Science and Technology,
29 Jiangjun Road, Nanjing 211106, China
mmisir@nuaa.edu.cn
[3] ASAP Research Group, School of Computer Science, University of Nottingham,
Jubilee Campus, Wollaton Road, Nottingham NG8 1BB, UK
ender.ozcan@nottingham.ac.uk

Abstract. Selection hyper-heuristics are high level search methodologies which control a set of low level heuristics while solving a given problem. Move acceptance is a crucial component of selection hyper-heuristics, deciding whether to accept or reject a new solution at each step during the search process. This study investigates group decision making strategies as ensemble methods exploiting the strengths of multiple move acceptance methods for improved performance. The empirical results indicate the success of the proposed methods across six combinatorial optimisation problems from a benchmark as well as an examination timetabling problem.

Keywords: Metaheuristic · Optimisation · Parameter control · Timetabling · Group decision making

1 Introduction

A selection hyper-heuristic is an iterative improvement oriented search method which embeds two key components; *heuristic selection* and *move acceptance* [3]. The heuristic selection method chooses and applies a heuristic from a set of low level heuristics to the solution in hand, producing a new one. Then the move acceptance method decides whether to accept or reject this solution. The modularity, use of machine learning techniques and utilisation of the *domain barrier* make hyper-heuristics more general search methodologies than the current techniques tailored for a particular domain are. A selection hyper-heuristic or its components can be reused on another problem domain without requiring any change. There is a growing number of studies on selection hyper-heuristics combining a range of simple heuristic selection and move acceptance methods [6,13].

© The Author(s) 2016
T. Czachórski et al. (Eds.): ISCIS 2016, CCIS 659, pp. 21–29, 2016.
DOI: 10.1007/978-3-319-47217-1_3

More on any type of hyper-heuristic, such as their components and application areas can be found in [3].

Hyper-heuristics Flexible Framework (HyFlex) [11] was proposed as a software platform for rapid development and testing of hyper-heuristics. HyFlex is implemented in Java along with six different problem domains: boolean satisfiability, bin-packing, permutation flow-shop, personnel scheduling, travelling salesman problem and vehicle routing problem. HyFlex was used in the first Cross-Domain Heuristic Search Challenge, CHeSC 2011 (http://www.asap.cs. nott.ac.uk/chesc2011/) to detect the best selection hyper-heuristic. Following the competition, the results from twenty competing selection hyper-heuristics across thirty problem instances (containing five instances from each HyFlex domain) and the description of their algorithms were provided at the competition web-page.

A recent theoretical study on selection hyper-heuristics in [10] showed that the mixing of simple move acceptance criteria could lead to an improved running-time complexity than using each move acceptance method standalone on some simple benchmark functions. In [1,8] different move acceptance criteria were used under an iterative two-stage framework which switches from one move acceptance to another at each stage. The previous work [2,13] indicates that the overall performance of a hyper-heuristic depends on the choice of selection hyper-heuristic components. This study extends the initial work in Özcan et al. [12] by applying and evaluating four group decision making strategies as ensemble methods using three different move acceptance methods in combination with seven heuristic selection methods on an examination timetabling problem [2]. The same selection hyper-heuristics are then tested on thirty problem instances from six different domains from the HyFlex benchmark.

2   Group Decision Making Selection Hyper-heuristics

An overview of heuristic selection and move acceptance methods as a part of the selection hyper-heuristics as well as the group decision making methods forming an ensemble of move acceptance used in this study is described in this section.

A range of simple heuristic selection methods were studied in [6]. *Simple Random* (SR) selects a heuristic at random at each decision point. *Random Descent* (RD) also selects a heuristic at random, and then applies it to the candidate solution as long as the solution is improved. *Random Permutation* (RP) generates a random permutation of heuristics and applies one heuristic at a time in that order. *Random Permutation Descent* (RPD) is based on the same RP strategy, however similar to RD, applies the same heuristic repeatedly until there is no more improvement. *Greedy* (GR) applies all low level heuristics to the current solution and selects the heuristic which generates the best improvement. *Choice Function* (CF) is an online learning heuristic selection method that scores each low level heuristic based on their utility value and selects the one with the highest score. A *Tabu Search* based hyper-heuristic (TABU) that maintains a tabu list of badly performing low level heuristics to disallow the selection of these heuristics was tested in [5].

This paper studies ensemble move acceptance methods combining them under a group decision making framework. Considering that a constituent move acceptance method returns either true (1) or false (0) at each decision point, Eq. 1 provides a general model for an ensemble of k methods. In this model, each move acceptance carries a certain *strength* (s_i) which adjusts its contribution towards a final acceptance decision.

$$\sum_{i=1}^{k} s_i \times D(M_i) \geq \alpha \qquad (1)$$

where M_i is the ith move acceptance (group member), $D(m)$ returns 1, if a solution is accepted by the move acceptance method m, and 0, if rejected.

In this study, we use group decision making strategies which make an accept/reject decision based on *authority*, *minority* and *majority* rules, namely G-OR (the move acceptance method which accepts the solution has the authority), G-AND (minority decides rejection), G-VOT and G-PVO (considers majority of the votes for the accept/reject decision). G-PVO probabilistically makes the accept/reject decisions. The probability that a new solution is accepted changes dynamically in proportional to the number of members that voted to the acceptance of the new solution. For instance, assuming 6 members in the group out of 10 move acceptance methods accepts a solution at a given step, then G-PVO accepts the solution with a probability of 60%. It is preferable in G-VOT to have an odd number of members for the group decision making move acceptance criteria, where none of the other strategies requires this. More formally, using Eq. 1, assuming k move acceptance methods, then for G-AND, G-OR and G-VOT, α is k, 0.5 and $k/2$, respectively, where all s_i values are set to 1. For G-PVO, α equals $k * r$, where r is uniform random number in $[0, 1]$, and s_i values equal $1/k$.

In this study, the heuristic selection methods in {SR, RD, RP, RPD, CF, GR, TABU} are paired with four group decision making move acceptance mechanisms {G-AND, G-OR, G-VOT, G-PVO}, generating twenty eight group decision making selection hyper-heuristics. From this point forward, a selection hyper-heuristic will be denoted as *"heuristic selection method"_"move acceptance method"*. For example, SR_G-AND denotes the selection hyper-heuristic using SR as the heuristic selection method and G-AND as the move acceptance method.

Each group decision making move acceptance ensemble tested in this study embeds three move acceptance methods: Improving and Equal (IE), Simulated Annealing (MC) and Great Deluge (GD). These group members are chosen to form the ensemble move acceptance due to their high performance reported in [13]. IE accepts all non-worsening moves and rejects the rest. Simulated Annealing [9] move acceptance criterion, denoted as MC in this paper, accepts all improving moves but the non-improving moves are accepted with a probabilistic formula, p_t, shown in Eq. 2.

$$p_t = e^{-\frac{\Delta f}{\Delta F(1 - \frac{t}{T})}} \qquad (2)$$

where Δf is the fitness change at time or step t, T is the time limit or the maximum number of steps and ΔF is an expected range for the maximum fitness change. GD acceptance criterion accepts all the improving moves but the non-improving moves are accepted if the objective value of the current solution is not worse than an expected value, named as level [7]. Equation 3 is used to update the threshold level τ_t at time or step t.

$$\tau_t = F + \Delta f \times (1 - \frac{t}{T})$$ (3)

where T is the time limit or the maximum number of steps, Δf is an expected range for the maximum fitness change and F is the final objective value.

3   Computational Experiments

Pentium IV 3 GHz LINUX machines having 2.00 GB memories are used during the experiments. Following the rules of CHeSC 2011, each trial is run for 10 nominal minutes with respect to the competition machine respecting the challenge rules. The group decision making selection hyper-heuristics are tested on an examination timetabling problem as formulated in [2] and the same termination criterion as in that study is used for the examination timetabling experiments to enable a fair performance comparison of solution methods. The GD and SA move acceptance methods use the same parameter settings as provided in [12].

Two sets of benchmarks are used for examination timetabling: Yeditepe [14,15] and Toronto benchmarks [4] consisting of eight and fourteen instances, respectively. The mean performance of each group decision making move acceptance method in a selection hyper-heuristic regardless of the heuristic selection method is compared to each other based on their ranks. The group decision making move acceptance methods are ranked from 1 to 4 for each problem instance and heuristic selection method from best to worst based on the mean cost over fifty runs. The approaches are assigned to different ranks if their performances vary in a statistically significant manner for a given instance. Otherwise, their performances are considered to be similar and an average rank is assigned to them all. A similar outcome is observed for the online performances of the

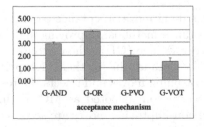

Fig. 1. Mean rank (and the standard deviation) of each group decision making move acceptance mechanism considering their average performance over all runs

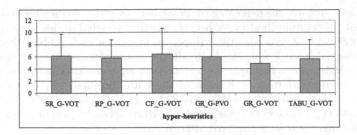

Fig. 2. Mean rank (and standard deviation) of the group decision making hyper-heuristics that generate statistically significant performance variance from the rest over all examination timetabling problems.

group decision making strategies as in the benchmark functions reported in [12]. G-VOT is the best acceptance mechanism based on the average rank over all the problems, while G-PVO, G-AND and G-OR follows it in that order, respectively as illustrated in Fig. 1.

Similarly, all twenty eight hyper-heuristics are ranked from 1 to 28 (best to worst) based on the best objective values obtained over fifty runs for each instance. The ranks are averaged/shared in case of a tie. Figure 2 illustrates the performance of six group decision making selection hyper-heuristics with a better mean performance that are significantly better as compared to the rest, from the best to the worst; GR_G-VOT, TABU_G-VOT, RP_G-VOT, GR_G-PVO, SR_G-VOT and CF_G-VOT.

Table 1 compares the average performances of the best six group decision making hyper-heuristics (see Fig. 2) to the best hyper-heuristic for each problem instance reported in [2]. Hyper-heuristics with multiple move acceptance methods under decision making models generated superior performance compared to the hyper-heuristics where each utilises a single move acceptance method. This performance variation is statistically significant within a confidence interval of 95 % based on the Wilcoxon signed-rank test. In eighteen out of the twenty one problems, hyper-heuristics with the majority rule voting as their acceptance criterion, namely G-VOT and G-PVO deliver the best performances. There is a tie between the simulated annealing based hyper-heuristics and group decision making hyper-heuristics for sta83 I and yue20013. It is also known that there is an optimal solution for yue20023 [15]. GR_G-PVO improves the average performance of CF_MC for yue20023, still, all the hyper-heuristics seem to get stuck at local optima while solving sta83 I, yue20013 and yue20023. Excluding yue20032, the group decision making hyper-heuristics improve the average performance of previous best hyper-heuristics by 30.7 % over all problem instances. RP_G-PVO delivers a similar average performance to CF_MC for yue20032, yet CF_MC is slightly better. Large improvements are observed for large problem instances, such as car91 I and car92 I. Overall, the experimental results confirm that group decision making hyper-heuristics have great potential.

Table 1. %*imp.* denotes the percentage improvement over the average best cost across fifty runs that the 'current' best hyper-heuristic(s) (investigated in this work) produces over the 'previous' best hyper-heuristic (reported in [2]) for each problem instance. If a hyper-heuristic delivers a statistically significant performance, it appears in the 'current' column. Bold entries highlight the best performing method. The hyper-heuristics that have a similar performance to the bold entry are displayed in parentheses. "+" indicates that all hyper-heuristics in {GR_G-VOT, TABU_G-VOT, RP_G-VOT, GR_G-PVO, SR_G-VOT, CF_G-VOT} has similar performance. "/" excludes the hyper-heuristic from this set that is displayed afterwards

instance	current	previous	%*imp.*
yue20011	**GR_G-VOT+**	SR_GD	20.84
yue20012	**RP_G-VOT+**	SR_GD	24.93
yue20013	**+**	**SR_MC**	0
yue20021	**TABU_G-VOT+**	SR_GD	17.97
yue20022	**GR_G-PVO**	CF_MC	3.97
yue20023	**GR_G-PVO**	CF_MC	1.97
yue20031	**GR_G-PVO** (GR_G-VOT, SR_G-VOT)	CF_MC	4.4
yue20032	n/a	**CF_MC**	n/a
car91 I	**GR_G-VOT+**	TABU_IE	81.37
car92 I	**GR_G-VOT+**/GR_G-PVO	TABU_IE	196.89
ear83 I	**GR_G-PVO** (GR_G-VOT)	CF_MC	1.1
hecs92 I	**GR_G-PVO** (GR_G-VOT, SR_G-VOT, TABU_G-VOT)	CF_MC	21.46
kfu93	**GR_G-VOT+**	SR_GD	30.88
lse91	**GR_G-PVO+**	CF_MC	13.38
pur93 I	**GR_G-PVO** (SR_G-VOT)	SR_IE	15.6
rye92	**TABU_G-VOT+**	CF_MC	41.67
sta83 I	**+**	**SR_MC**	0
tre92	**GR_G-VOT+**	SR_GD	92.93
uta92 I	**GR_G-VOT+**/GR_G-PVO	TABU_IE	36.36
ute92	**GR_G-PVO**	**CF_MC**	0
yor83 I	**GR_G-PVO+**	CF_MC	9.01

The twenty eight hyper-heuristics are implemented as an extension to HyFlex to check their level of generality across the CHeSC 2011 problem domains. Each experiment is repeated thirty one times following the competition rules. All hyper-heuristics are ranked using the Formula 1 scoring system. The best hyper-heuristic obtaining the best median objective value over all runs for each instance gets 10 points, the second one gets 8, and then 6, 5, 4, 3, 2, 1 and the rest gets zero point. These points are accumulated over all instances across all domains forming the final score for each hyper-heuristic.

Firstly, performance of all group decision making hyper-heuristics are compared to each other. Figure 3 summarises the results including top twelve out of twenty eight approaches. In the overall, CF_G-OR, CF_G-VOT and TABU_G-VOT are the top three group decision making methods, while GR_G-AND and

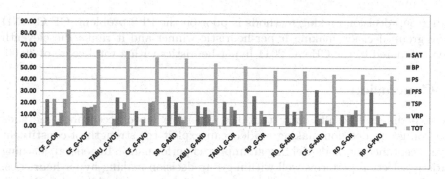

Fig. 3. Median performance comparisons between different group decision making hyper-heuristics based on their Formula 1 scores.

GR_G-OR are the worst. RP_G-PVO, CF_G-AND, CF_G-OR, TABU_G-VOT, CF_G-PVO and CF_G-OR perform the best on boolean satisfiability (SAT), bin-packing (BP), personnel scheduling (PS), permutation flow-shop (PFS), travelling salesman (TSP) and vehicle routing problems (VRP), respectively. Table 2 summarises the ranking of those six group decision making hyper-heuristics and all competing hyper-heuristics at CHeSC 2011, including the top ranking method, denoted as AdapHH. The top ten ranking hyper-heuristics from the competition remains in their positions and group decision making methods perform relatively poor. CF_G-AND is the third best approach for BP. TABU_G-VOT comes sixth for PS. TABU_G-VOT, CF_G-AND and CF_G-VOT score better than the CHeSC 2011 winner for the same problem. CF_G-OR is the best among the group decision making methods for SAT, ranking the eighth. The best group decision making hyper-heuristic for TSP, i.e. CF_G-OR, takes the ninth place. For VRP, CF_G-VOT as the best hyper-heuristic with group decision making is the sixth best approach among the CHeSC 2011 competitors. However, its performance on VRP is still better than the winning approach. The performance

Table 2. Ranking of selected group decision making hyper-heuristics to the CHeSC 2011 competitors based on Formula 1

Rank	HH	Total	SAT	BP	PS	PFS	TSP	VRP
1	AdapHH	170.00	33.75	43.00	6.00	37.00	40.25	10.00
7	HAHA	65.75	31.75	0.00	19.50	3.50	0.00	11.00
11	CF_G-AND	39.00	0.00	25.00	10.00	0.00	0.00	4.00
14	CF_G-OR	27.50	9.50	0.00	2.00	0.00	8.00	8.00
15	CF_G-VOT	23.50	0.00	0.00	8.50	0.00	4.00	11.00
20	CF_G-PVO	16.14	0.14	0.00	1.00	0.00	6.00	9.00
22	TABU_G-VOT	11.50	0.00	0.00	11.50	0.00	0.00	0.00
23	RP_G-PVO	7.00	6.50	0.00	0.50	0.00	0.00	0.00

of all group decision making methods is poor on the PFS problem. CF_G-AND is the group decision making hyper-heuristic winner and it ranks the eleventh when compared to the CHeSC 2011 hyper-heuristics with a total score of 39.00.

4   Conclusion

The experimental results show that the ensemble move acceptance methods based on group decision making models can exploit the strength of constituent move acceptance methods yielding an improved performance. In general, learning heuristic selection performs well within group decision making hyper-heuristics. Considering their performance over the examination timetabling benchmark problems, Greedy performs the best as a heuristic selection method. Combining multiple move acceptance methods using a majority rule improves the performance of Greedy as compared to using a single move acceptance method. On the other side, CF outperforms other standard heuristic selection schemes on the CHeSC 2011 benchmark, performing reasonably well in combination with AND-operator group decision making move acceptance. The proposed ensemble move acceptance methods enable the use of the existing move acceptance methods and do not introduce any extra parameters other than the constituent methods have. Discovering the best choice of move acceptance methods in the ensemble as well as their weights is left as a future work. More interestingly, new adaptive ensemble move acceptance methods, which are capable of adjusting the weight/strength of each constituent move acceptance during the search process, can be designed for improved cross domain performance.

References

1. Asta, S., Özcan, E.: A tensor-based selection hyper-heuristic for cross-domain heuristic search. Inf. Sci. **299**, 412–432 (2015)
2. Bilgin, B., Özcan, E., Korkmaz, E.E.: An experimental study on hyper-heuristics and exam timetabling. In: Burke, E.K., Rudová, H. (eds.) PATAT 2006. LNCS, vol. 3867, pp. 394–412. Springer, Heidelberg (2007). doi:10.1007/978-3-540-77345-0_25
3. Burke, E.K., Gendreau, M., Hyde, M., Kendall, G., Ochoa, G., Özcan, E., Qu, R.: Hyper-heuristics: A survey of the state of the art. J. Oper. Res. Soc. **64**(12), 1695–1724 (2013)

4. Carter, M.W., Laporte, G., Lee, S.Y.: Examination timetabling: algorithmic strategies and applications. J. Oper. Res. Soc. **47**(3), 373–383 (1996)
5. Cowling, P., Chakhlevitch, K.: Hyperheuristics for managing a large collection of low level heuristics to schedule personnel. In: IEEE Congress on Evolutionary Computation, pp. 1214–1221 (2003)
6. Cowling, P.I., Kendall, G., Soubeiga, E.: A hyperheuristic approach to scheduling a sales summit. In: Burke, E., Erben, W. (eds.) PATAT 2000. LNCS, vol. 2079, p. 176. Springer, Heidelberg (2001)
7. Kendall, G., Mohamad, M.: Channel assignment optimisation using a hyperheuristic. In: IEEE Conference on Cybernetic and Intelligent Systems, pp. 790–795, 1–3 December 2004
8. Kheiri, A., Özcan, E.: An iterated multi-stage selection hyper-heuristic. Eur. J. Oper. Res. **250**(1), 77–90 (2016)
9. Kirkpatrick, S., Gelatt, C.D., Vecchi, M.P.: Optimization by simulated annealing. Science **220**, 671–680 (1983)
10. Lehre, P.K., Özcan, E.: A runtime analysis of simple hyper-heuristics: to mix or not to mix operators. In: Workshop on Foundations of Genetic Algorithms XII, pp. 97–104 (2013)
11. Ochoa, G., et al.: HyFlex: a benchmark framework for cross-domain heuristic search. In: Hao, J.-K., Middendorf, M. (eds.) EvoCOP 2012. LNCS, vol. 7245, pp. 136–147. Springer, Heidelberg (2012)
12. Özcan, E., Misir, M., Kheiri, A.: Group decision making hyper-heuristics for function optimisation. In: The 13th UK Workshop on Computational Intelligence, pp. 327–333, September 2013
13. Özcan, E., Bilgin, B., Korkmaz, E.E.: A comprehensive analysis of hyper-heuristics. Intell. Data Anal. **12**(1), 3–23 (2008)
14. Özcan, E., Ersoy, E.: Final exam scheduler - fes. In: Corne, D., Michalewicz, Z., McKay, B., Eiben, G., Fogel, D., Fonseca, C., Greenwood, G., Raidl, G., Tan, K.C., Zalzala, A. (eds.) IEEE Congress on Evolutionary Computation, pp. 1356–1363 (2005)
15. Parkes, A.J., Özcan, E.: Properties of yeditepe examination timetabling benchmark instances. In: PATAT VIII, pp. 531–534 (2010)

Extending Static Code Analysis with Application-Specific Rules by Analyzing Runtime Execution Traces

Ersin Ersoy[1] and Hasan Sözer[2]($\boxtimes$)

[1] Turkcell Technology, İstanbul, Turkey
ersin.ersoy@turkcell.com.tr
[2] Ozyegin University, İstanbul, Turkey
hasan.sozer@ozyegin.edu.tr

Abstract. Static analysis tools cannot detect violations of application-specific rules. They can be extended with specialized checkers that implement the verification of these rules. However, such rules are usually not documented explicitly. Moreover, the implementation of specialized checkers is a manual process that requires expertise. In this work, application-specific programming rules are automatically extracted from execution traces collected at runtime. These traces are analyzed offline to identify programming rules. Then, specialized checkers for these rules are introduced as extensions to a static analysis tool so that their violations can be checked throughout the source code. We implemented our approach for Java programs, considering 3 types of faults. We performed an evaluation with an industrial case study from the telecommunications domain. We were able to detect real faults with checkers that were generated based on the analysis of execution logs.

1 Introduction

Static code analysis tools (SCAT) can detect the violation of programming rules by checking (violation of) patterns throughout the source code [1]. The detected violations are reported in the form of a list of alerts. Although SCAT have been successfully utilized in the industry [7,8,15], they have limitations as well. It is very hard or undecidable to show whether an execution path is feasible or infeasible without the runtime context information [11]. As a result, some faults might be missed. SCAT also fall short to detect the violation of application-specific rules [3]. For example, it might be necessary to check some of the arguments and/or return values before/after certain method calls. SCAT do not consider such application-specific rules by default.

One can extend SCAT with specialized checkers to detect the violation of application-specific rules [3]. However, the implementation of specialized checkers is a manual process that requires expertise. In fact, state-of-the-art SCAT provide special extension mechanisms for defining new rules, which can be then checked by these tools. Yet, such rules have to be defined manually and they are usually not documented explicitly or formally.

© The Author(s) 2016
T. Czachórski et al. (Eds.): ISCIS 2016, CCIS 659, pp. 30–38, 2016.
DOI: 10.1007/978-3-319-47217-1_4

In this paper, we introduce an approach for extending SCAT, in which specialized checkers are generated automatically. Our approach employs offline analysis of execution traces collected at runtime. These traces comprise a set of encountered errors. The source code is analyzed to identify faults that are the root causes of these errors. One could consider just to fix these faults without systematically and formally documenting them. However, instances of the same fault can exist at other places in the source code. It might also be possible that the same fault is introduced again later on. Therefore, it is important to capture this information and systematically check for the identified faults in the overall source code regularly. In our approach, programming rules are inferred to prevent these pitfalls. Specialized checkers are automatically generated for these rules and they are introduced as extensions to SCAT. The extended SCAT can detect the violation of the inferred rules throughout the source code.

We performed an evaluation with an industrial case study from the telecommunications domain. We captured the execution logs of a previous version of a large scale system implemented in Java. A number of recorded errors are analyzed for 3 types of errors and the corresponding faults are identified. We generated rules and specialized checkers for these faults, which were already fixed. The SCAT that is employed in the company is extended with these checkers. Then, we were able to detect several new instances of the identified faults that had to be fixed.

The remainder of this paper is organized as follows. The following section summarizes the related studies. We present the overall approach in Sect. 3. The approach is illustrated in Sect. 4, in the context of the industrial case study. Finally, in Sect. 5, we conclude the paper.

2   Related Work

There have been studies for automatically deriving programming rules based on frequently used code patterns [4,5]. Hereby, pattern recognition, data mining and heuristic algorithms are used for analyzing the program source code and detecting potential rules. Then, the source code is analyzed again to detect inconsistencies with respect to these rules. These studies utilize only (models of) the source code to infer programming rules. They do not make use of runtime execution traces.

There are studies [2,14] that make use of the analysis of previously fixed bugs to derive application-specific programming rules. However, programmers have to define the rules applied to fix these bugs. Hence, they rely on manual analysis. In addition, they do not exploit any information collected during runtime execution.

There exist a few approaches [9,10,13] that exploit dynamic analysis and runtime execution traces. DynaMine [9] uses dynamic analysis for validating programming rules that are actually derived by mining the revision history. Another approach [13] relies on the analysis of console logs to detect anomalies [13]; however, deriving rules for preventing these anomalies was out of the scope of the study. Daikon [10] derives likely invariants of a program by means of dynamic analysis. However, Daikon focuses on numerical properties of variables as system constraints rather than bug patterns that can represent a wider range of bug types.

We have previously introduced an approach to generate runtime monitors based on SCAT alerts [12] These monitors identify alerts, which do not actually cause any failures at runtime. Then, filters are automatically generated for SCAT to supress these alerts. Hence, the goal is to reduce false positives and increase precision. In this work, we aim at reducing false negatives by detecting more faults as a result of checking application-specific rules. As such, the goal of the approach proposed in this paper is to increase recall instead.

3   Generating Rules from Execution Traces

Our approach takes runtime execution traces of a system as input. These traces should comprise the set of errors encountered and the set of software modules involved. The output is a set of checkers that are provided as extensions to SCAT. These checkers detect instances of faults that are the root causes of the logged errors. To be able to identify these faults and to generate the corresponding checkers, a library of analysis procedures and a library of checker templates are utilized, respectively. The scope of these libraries define the set of error and fault types that can be considered by the approach.

The overall process is depicted in Fig. 1, which involves 4 steps. First, *Log Parser* takes runtime logs as input, parses these logs, and generates the list of errors recorded together with the related modules and events (1). Then, this list is provided to *Root Cause Analyzer*, which analyzes the source code to identify the cause of the error by utilizing a set of predefined analysis procedures (2). For instance, if a null pointer reference error is detected at runtime, the corresponding analysis procedure locates the corresponding object and its last definition before the error. Let's assume that such an object was defined as the return value of a method call. Then, a rule is inferred, imposing that the return value of that particular method must be checked before use. The list of such rules are provided to *Checker Generator*, which uses a library of predefined templates to generate

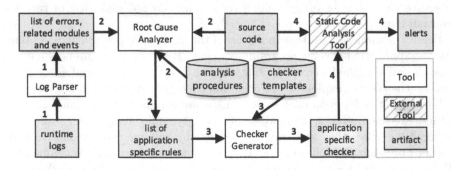

Fig. 1. The overall process.

a specialized checker for each rule (3). The generated checkers are included as extensions to SCAT, which applies them to the source code and reports alerts in case violations are flagged (4).

The overall process is automated; however, it relies on a set of predefined analysis procedures and checker templates. One analysis procedure should be defined for each error type and one checker template should be defined for each rule type. The set of rules and error types are open-ended in principle and they can be extended when needed. Currently, we consider the following types of errors and programming rules that are parametrized with respect to the involved method and argument names.

- *java.lang.IndexOutOfBoundsException*: The arguments of a method must be checked for boundary values before the method call, e.g., *if(x < MAX) m(x);*
- *java.lang.NullPointerException*: The return value of a method must be checked for null reference, e.g., *r = m(x); if(r != null) {...}* or *if(r == null) {...}*
- *org.hibernate.LazyInitializationException*: The JPA Entity[1] should be initialized at a transactional level (when persistence context is alive) before being used at a non-transactional level, e.g., object *a* is a JPA Entity with *LAZY* fetch type and it is an aggregate within object *b*. Then, *a* must be fecthed from the database when *b* is being initialized, for a possible access after the persistant context is lost.

In the following, we explain the steps of the approach in more detail with a running example. Then, in Sect. 4, we illustrate the application of the approach in the context of an industrial case study[2].

Analysis of Execution Logs: The first step of our approach involves the analysis of execution logs. In our case study, we had to utilize existing log files of a legacy system. Therefore, *Log Parser* is implemented as a dedicated parser for these files. However, it can be replaced with any parser to be able to process log files in other formats as well. Our approach is agnostic to the log file structure as long as the following information can be derived: *(i)* Sequence of events and in particular, encountered errors; *(ii)* The types of encountered errors; *(iii)* The location of the encountered errors in the source code, i.e., package, class, method name, line number. Even standard Java exception reports include such information together with a detailed stack trace. Hence, existing instrumentation and logging tools can be employed to obtain the necessary information. *Log Parser* is parametric with respect to the focused error types and modules of the system. We can filter out some error types or modules that are deemed irrelevant or uncritical.

---

[1] A JPA (Java Persistence API) entity is a POJO (Plain Old Java Object) class, which has the ability to represent objects in a database. They can be reached within a persistent context.

[2] Currently our toolset works on software systems written in Java. In principle, the approach can be instantiated for different programming languages/environments. Our design and implementation choices were driven by the needs and the context of the industrial case.

Root Cause Analysis: Once *Log Parser* retrieves the relevant error records together with their context information, it provides them to *Root Cause Analyzer*. This tool performs two main tasks: *(i)* finding the root cause of the error, *(ii)* determining whether this root cause is application-specific or not. We are not interested in generic errors. Hence, it is important to be sure that the root cause of the error is application-specific. For instance, consider the code snippet in Listing 1.1. When executed, it causes a *java.lang.NullPointerException*; however, *Root Cause Analyzer* ignores this error because, the cause of the error is an object that is simply left unitialized. This is a generic error.

Listing 1.1. An sample code snippet for a generic error that is ignored by *Root Cause Analyzer*.

```
1   static Report aReport;
2   public static void print() { System.out.println(aReport); }
```

If the null value is obtained from a specific method in the application, then such an error is deemed relevant (See Listing 1.2). That means, the return value of the corresponding method (e.g., *getServiceReport*) must be always checked before use. This is a type of rule that is determined by *Root Cause Analyzer*.

Listing 1.2. A possible application-specific error that is considered by *Root Cause Analyzer*.

```
1   static Report aReport = getServiceReport();
2   public static void print() { System.out.println(aReport); }
```

Root Cause Analyzer employs a set of predefined analysis procedures that are coupled with error types. For example, the analysis procedure applied for null pointer exceptions is listed in Algorithm 1. Hereby, the use of the object that caused a null pointer exception is located as the first step. Second, the reaching definition is found for this use of the object. If this definition is performed with a method call, the procedure checks where the method is defined. If the method is defined within the application, then a rule is reported for checking the return value of this method.

Root Cause Analyzer provides the type of rule to be applied and the parameters of the rule (e.g., name of the method, of which return value must be checked) to *Checker Generator* so that a specialized checker can be created.

Algorithm 1. Root cause analysis procedure applied for null pointer exceptions.

1: $u \leftarrow$ use of object that causes the exception
2: $d \leftarrow$ reaching definition for u
3: **if** $\exists$ method m as part of d **then**
4:     $p \leftarrow$ package of m
5:     **if** $p \in$ application packages **then**
6:         $reportRule(\text{RETURNVALCHECK}, m)$
7:     **end if**
8: **end if**

Generation of Specialized Checkers: Most SCAT are extensible; they provide application programming interfaces (API) for implementing custom checkers. *Checker Generator* generates specialized checkers by utilizing PMD[3] as SCAT. PMD uses JavaCC[4] to parse the source code and generate its abstract syntax tree (AST). This AST can be traversed with its Java API to define specialized checkers for custom rules. These checkers should conform to the Visitor design pattern [6]. Each checker is basically defined as an extension of an abstract class, namely, *AbstractJavaRule*. The *visit* method that is inherited from this class must be overwritten to implement the custom check. This method takes two arguments: *(i) node* of type *ASTMethodDeclaration* and *(ii) data* of type *Object*. The return value is of type *Object*. This visitor method is called by PMD for each AST node (e.g., method).

Checker generation is performed based on parametrized templates. We defined a template for each rule type. Each template extends the *AbstractJavaRule* class and overwrites the necessary visitor methods. A checker is generated by instantiating the corresponding template by assigning concrete values to its parameters. For instance, consider a specialized checker that enforces the handling of possible null references returned from a method in the application. The corresponding pseudo code that is implemented with PMD is listed in Algorithm 2. Hereby, all variable declarations are obtained as a set (V at Line 1). For each of these declarations (v), the node ID (vid) is obtained (Line 3). The name of the method call (m) is also obtained, assuming that the declaration involves a method call (Line 4). If there indeed exists such a method call and if the name of the method matches the expected name (i.e., $METHOD$), then an additional check is performed (*isNullCheckPerformed* at Line 6). This check traverses the AST starting from the node with id vid and searches for control statements that compare the corresponding variable (v) with respect to null (i.e., *if(v != null) {...}* or *if(v == null) {...}*). If there is no such a control statement before the use of the variable, then a violation of the rule is registered (Line 8).

Checker Generator generates specialized checkers by instantiating the corresponding template with the parameters (e.g., $METHOD$) provided by *Root Cause Analyzer*. Hence, multiple checkers can be generated based on the same rule type.

Extension of Static Code Analysis Tool: PMD is extended with the custom checkers generated by *Checker Generator* and it is executed by Sonar[5] version 4.0. The extension is performed in two steps: *(i)* adding a jar file that includes the custom checker, and *(ii)* extending the XML configuration file for rule definition. The jar file basically contains an instantiation of a checker code template. The rule regarding the introduced checker is defined in the XML configuration file by a new entry pointing at this jar file. It also specifies the *name*, *message* and *description* of the rule, which are displayed to the user as part of the listed alerts, when violations are detected.

---

[3] http://pmd.sourceforge.net/.

[4] https://javacc.java.net/.

[5] http://www.sonarqube.org/.

Algorithm 2. visit method of a specialized checker for a custom rule, i.e., *handle possible null pointer after calling the method.*

1: $V = getChildrenOfType(ASTVariableDeclarator)$;
2: **for all** $v \in V$ **do**
3:     $vid = v.getID()$;
4:     $m = v.getMethodCall()$;
5:     **if** $m! = null$ & $m.name == METHOD$ **then**
6:         $isChecked = isNullCheckPerformed(vid)$
7:         **if** $!isChecked$ **then**
8:             $addViolation(vid)$
9:         **end if**
10:     **end if**
11: **end for**

---

4   Industrial Case Study

We performed a case study on a Sales Force Automation system maintained by Turkcell[6]. The system comprises more than 200 KLOC. It is operational since 2013, serving 2000 users. We downloaded all the log files regarding a previous version of this system. *Log Parser* identified an error in these files. The corresponding source code snippet is listed in Listing 1.3, where the object *opty* turns out to be null. Then, *Root Cause Analyzer* located the point in the source code, where this object was last defined (Line 1). The definition is coming from a method call, i.e., *templateDao.find(Opty.class, optyNo);*. This method creates and returns an object by utilizing information from a database; it returns null if the required information cannot be found.

Listing 1.3. The code snippet corresponding to the logged error.

```
1  Opty opty = templateDao.find(Opty.class, optyNo);
2  if (opty.getCoptycategory().equals(...)) { ... }
```

Then, an application-specific rule is inferred as: the return value of the method *find* must be checked for null references before use. A specialized checker is automatically generated based on this rule. It checks the whole code base and searches for initialized objects using the return value of the method *find* without a null reference check. As the last step, Sonar is extended with the specialized checker.

After the extension, 25 additional alerts were generated. All the alerts were true positives and the corresponding code locations really required to be fixed. In fact, we have seen that 3 of these locations caused errors afterwards and they were fixed in a later version of the source code. If our approach were applied and all the reported alerts were addressed, these errors would not occur at all. As a result, 25 real faults were detected with specialized checkers and 3 of them were

---

[6] http://www.turkcell.com.tr.

activated during operational time. This result shows the importance and high potential of information collected at runtime as a source for improving recall in static analysis.

5   Conclusion

In this work, we extracted application-specific programming rules by analyzing logged errors. We automatically generated specialized checkers for these rules as part of a static code analysis tool. Then, the tool can check for potential instances of the same type of error throughout the source code. We conducted an industrial case study from the telecommunications domain. We were able to detect real faults, which had to be fixed later on. In the future, we plan to extend our approach to cover more than 3 types of errors and rules. We also plan to conduct more case studies.

Acknowledgments. This work is supported by The Scientific and Research Council of Turkey (113E548).

References

1. Johnson, B., et al.: Why don't software developers use static analysis tools to find bugs?. In: Proceedings of the 35th International Conference on Software Engineering, pp. 672–681 (2013)
2. Sun, B., et al.: Automated support for propagating bug fixes. In: Proceedings of the 19th International Symposium on Software Reliability Engineering, pp. 187–196 (2008)
3. Sun, B., et al.: Extending static analysis by mining project-specific rules. In: Proceedings of the 34th International Conference on Software Engineering, pp. 1054–1063 (2012)
4. Chang, R., Podgurski, A.: Discovering programming rules and violations by mining interprocedural dependences. J. Softw. Mainten. Evol. Res. Pract. **24**, 51–66 (2011)
5. Chang, R., Podgurski, A., Yang, J.: Discovering neglected conditions in software by mining dependence graphs. IEEE Trans. Softw. Eng. **34**(5), 579–596 (2008)
6. Gamma, E., Helm, R., Johnson, R., Vlissides, J.: Design Patterns: Elements of Reusable Object-oriented Software. Addison-Wesley, Boston (1995)

7. Zheng, J., et al.: On the value of static analysis for fault detection in software. IEEE Trans. Softw. Eng. **32**(4), 240–253 (2006)
8. Krishnan, R., Nadworny, M., Bharill, N.: Static analysis tools for security checking in code at motorola. ACM SIG Ada Lett. **28**(1), 76–82 (2008)
9. Livshits, B., Zimmerman, T.: Dynamine: finding common error patterns by mining software revision histories. SIGSOFT Softw. Eng. Not. **30**, 296–305 (2005)
10. Ernst, M.D., et al.: The Daikon system for dynamic detection of likely invariants. Sci. Comput. Program. **69**(1–3), 35–45 (2007)
11. Ayewah, N., et al.: Using static analysis to find bugs. IEEE Softw. **25**(5), 22–29 (2008)
12. Sozer, H.: Integrated static code analysis and runtime verification. Softw. Pract. Exp. **45**(10), 1359–1373 (2015)
13. Xu, W., et al.: Detecting large-scale system problems by mining console logs. In: Proceedings of the 22nd ACM Symposium on Operating Systems Principles, pp. 117–132 (2009)
14. Williams, C., Holingsworth, J.: Automatic mining of source code repositories to improve bug finding techniques. IEEE Trans. Softw. Eng. **31**, 466–480 (2005)
15. Yuksel, U., Sozer, H.: Automated classification of static code analysis alerts: a case study. In: Proceedings of the 29th IEEE International Conference on Software Maintenance, pp. 532–535 (2013)

The Random Neural Network Applied to an Intelligent Search Assistant

Will Serrano[✉]

Intelligent Systems and Networks Group Electrical and Electronic Engineering,
Imperial College London, London, UK
g.serrano11@imperial.ac.uk

Abstract. Users can not guarantee the results they obtain from Web search engines are exhaustive, or that they actually respond to their needs. Search results are influenced by the users' own ambiguity in formulating their requests or queries as well as by the commercial interest of Web search engines and Internet users that want to reach a wider audience. This paper presents an Intelligent Search Assistant (ISA) based on a Random Neural Network that acts as the interface between users and search engines to present data to users in a manner that reflects their actual needs or their observed or stated preferences. Our ISA tracks the user's preferences and makes a selection on the output of one or more search engines using the preferences that it has learned. We also introduce a "relevance metric" to compare the performance of our Intelligent Search Assistant against a few search engines, showing that it provides better performance.

Keywords: Intelligent search assistant · World wide web · Random neural network · Web search · Search engines

1 Introduction

Web Search Engines have been used as the direct connection between users and the information or products sought in the Internet. Search results are influenced by a commercial interest as well as by the users' own ambiguity in formulating their requests or queries. Ranking algorithms are essential in Web search as they decide the relevance; they make information visible or hidden to customers or users. Under this model, Web search engines or recommender systems can be tempted to artificially rank results from some specific businesses for a fee whereas also authors or business can be tempted to manipulate ranking algorithms by "optimizing" the presentation of their work or products. The main consequence is that irrelevant results may be shown on top positions and relevant ones "hidden" at the very bottom of the search list.

In order to address the presented search issues; this paper proposes an Intelligent Search Assistant (ISA) that acts as an interface between an individual user's query and the different search engines. Our ISA acquires a query from the user and retrieves results from one or various search engines assigning one neuron per each Web result dimension. The result relevance is calculated by applying our innovative cost function based on the division of a query into a multidimensional vector weighting its dimension

© The Author(s) 2016
T. Czachórski et al. (Eds.): ISCIS 2016, CCIS 659, pp. 39–51, 2016.
DOI: 10.1007/978-3-319-47217-1_5

terms with different relevance parameters. Our ISA adapts and learns the perceived user's interest and reorders the retrieved snippets based in our dimension relevant centre point. Our ISA learns result relevance on an iterative process where the user evaluates directly the listed results. We evaluate and compare its performance against other search engines with a new proposed quality definition, which combines both relevance and rank. We have also included two learning algorithms; Gradient Descent learns the centre of relevant dimensions and Reinforcement Learning updates the network weights based on rewarding relevant dimensions and punishing irrelevant ones. We have validated our ISA against other Web search engines using travel services and open user queries. We have also analysed the Gradient Descent and Reinforcement Learning algorithms based on result relevance and learning speed.

We describe the application of neural networks in Web search in Sect. 2. We define our Intelligent Search Assistant mathematical model in Sect. 3 and we have validated it against other Web search engines in Sect. 4. Finally, we present our conclusions in Sect. 5.

2   Related Work

Neural networks have been already applied in the World Wide Web as a mechanism of adaptation to users' interest in order to provide relevant answers. Wang et al. [1] use a back propagation neural network with its input nodes corresponding to an specific quantified user profile and one output node which it is the a probability the user would consider the Web page relevant. Boyan et al. [2] use reinforcement learning to rank Web pages using their HTML properties and hyperlink connections between them. Shu et al. [3] retrieve results from different Web search engines and train the network following the assumption that a result in a top position would be relevant. Burgues et al. [4] define RankNet which uses neural networks to evaluate Web sites by training the neural network based on query-document pairs. Bermejo et al. [5] use a similar approach to our proposal, the allocation of one neuron per Web search result, however the main difference is that the network is trained to cluster results by meaning. Scarselli et al. [6] use a neural network by assigning a neuron to each Web page; they create a graph where the neural links are the equivalent of the hyperlinks.

3   The Intelligent Search Assistant Model

The search assistant we design is based on the Random Neural Network (RNN) [7–9, 19]. This is a spiking recurrent stochastic model for neural networks. Its main analytical properties are the "product form" and the existence of the unique network steady state solution. The RNN represents more closely how signals are transmitted in many biological neural networks where they actual travel as spikes or impulses, rather than as analogue signal levels. It has been used in different applications including network routing with cognitive packet networks [10], search for exit routes for evacuees in emergency situations [11, 12], pattern based search for specific objects [13], video compression [14], and image texture learning and generation [15].

3.1   Search Model

In the case of our own application of the RNN, the search for information or for some meaning requires us to specify: an M-dimensional universe of X entities or ideas to be searched, a high level query that specifies the N-properties or concepts requested by a user and a method that searches and selects Y entities from the universe showing the first Z results to user according to an algorithm or rule. Each entity or concept in the universe is distinct from the others in some recognizable way; for instance two entities may be different just in the date or time-stamp that characterizes the time when they were last stored or in the ownership or origin of the entities. On the other hand, we consider concepts to be distinct if they contain any different meaning, even though if they are identical with respect to a user's query.

We consider that the universe which we are searching within as a relation U that consists of a set of X M-tuples, $U = \{v_1, v_2 \dots v_X\}$, where $v_i = (l_{i1}, l_{i2} \dots l_{iM})$ and li are the M different attributes for $i = 1, 2..X$. The relation U is a very large relation consisting on $M \gg N$ attributes. The important concept in the development of this paper is a query can be defined as $R_t(n(t)) = (R_t(1), R_t(2), \dots, R_t(n(t)))$ where n(t) is a variable N-dimension attribute vector with $1 < N < M$ and t is the search iteration being $t > 0$; n (t) is variable so that attributes can be added or removed based on their relevance as the search progresses, i.e. as t increases. Each $R_t(n(t))$ takes its values from the attributes within the domain $D(n(t))$, where D is the corresponding domain that forms the universe U. Thus $D(n(t))$ is a set of properties or meanings based in words or integers, but also words in another language, or a set of icons, images or sounds.

The answer A to the query $R_t(n(t))$ is a set of Y M-tuples $A = \{v_1, v_2 \dots v_Y\}$ where $v_o = (l_{o1}, l_{o2} \dots l_{oM})$ and lo are the M different attributes for $o = 1, 2..Y$. Our Intelligent Search Assistant only shows to the user the first set of Z tuples that have the highest neuron potentials among the set of Y tuples. The neuron potential that represents the relevance of each M-tuple v_o is calculated at each t iteration. The user or the high level query itself is limited mainly by two main factors: the user's lack of information about all the attributes that form the universe U of entities and ideas, or the user's lack of precise knowledge about what he is looking for.

3.2   Result Cost Function

We consider the universe U is formed of the entire results that can be searched. We assign each result provided by a search engine to an M-tuple v_o of the answer set A. We calculate the result relevance based on a cost function described within this section. The query $R_t(n(t))$ is a variable N-dimension vector that specifies the attributes the user consider relevant. The number of dimensions of the attribute vector n(t) varies as the iteration t increases. Our Intelligent Search Assistant associates an M-tuple v_o to each result provided by the Search Engine creating an answer set A of Y M-tuples. Search Engines select their results from the universe U. We apply our cost function to each result or M-tuple v_o from the answer set A of Y M-tuples. We consider each v_o as a M-dimensional vector. The cost function is firstly calculated based on the relevant N attributes the user introduced on the High Level Query, $R_1(n(1))$ within the domain

$D(n(1))$ however, as the search progresses, $R_t(n(t))$, attributes may be added or removed based on the perceived relevance within the domain $D'(n(t))$. We calculate the overall Result Score, RS, by measuring the relationship between the values of its different attributes:

$$RS = RV * HW \qquad (1)$$

where RV is the Result Value which measures the result relevance and HW the Homogeneity Weight. The Homogeneity Weight (HW) rewards results that have relevance or scores dispersed along their attributes. This parameter is also based on the idea that the first dimensions or attributes of the user query $R_t(n(t))$ are more important than the last ones:

$$HW = \frac{\sum_{n=1}^{N} HF[n]}{N} \qquad (2)$$

where HF[n], homogeneity factor, is a N-dimension vector associated to the result and n is the attribute index from the query $R_t(n(t))$:

$$HF[n] = \begin{vmatrix} \frac{N-n}{N} & \text{if } SD[n] > 0 \\ 0 & \text{if } SD[n] = 0 \end{vmatrix} \qquad (3)$$

We define Score Dimension SD[n] as a N-dimension vector that represents the attribute values of each result or M-tuple v_o in relation with the query $R_t(n(t))$. The Result Value (RV) is the sum of each dimension individual score:

$$RV = \sum_{n=1}^{N} SD[n] \qquad (4)$$

where n is the attribute index from the query $R_t(n(t))$. Each dimension of the Score Dimension vector SD[n] is calculated independently for each n-attribute value that forms the query $R_t(n(t))$:

$$SD[n] = S * PPW * RPW * DPW \qquad (5)$$

We consider only three different types of domains of interest: words, numbers (as for dates and times) and prices. S is the score calculated depending if the domain of the attribute is a word (WS), number (NS) or price (PS). If the domain $D(n)$ is a word, our ISA calculates the score Word Score (WS) following the formula:

$$S = \frac{WR}{NW} \qquad (6)$$

where the value of WR is 1 if the word of the n-attribute of the query $R_t(n(t))$ is contained in the search result or 0 otherwise. NW is the number of words in the search result. If the domain D(n) is a number, our ISA selects the best Number Score (NS) from the numbers they are contained within the search result that maximizes the cost function:

$$S = \frac{\left(1 - \left(\frac{|DV - RV|}{|DV| + |RV|}\right)\right)}{NN} \tag{7}$$

where DV is the value of the n-attribute of the query $R_t(n(t))$, RV is the value of a number in the result and NN is the total number of numbers in the result. If the domain D(n) is a price, our ISA chooses the best Price Score (PS) from the prices in the result that maximizes the cost function:

$$S = \frac{\left(\frac{DV}{RV}\right)}{NP} \tag{8}$$

where DV is value of the n-attribute of the query $R_t(n(t))$, RV is the value of a price in the result and NP is the total number of prices in the result. We penalize if the search result provides unnecessary information by dividing the score by the total amount of elements in the Web result. The dimension Score Dimension vector, SD[n] is weighted according to different relevance factors:

$$SD[n] = S * PPW * RPW * DPW \tag{9}$$

The Position Parameter Weight (PPW) is based on the idea that an attribute value shown within the first positions of the search result is more relevant than if it is shown at the final:

$$PPW = \frac{NC - DVP}{NC} \tag{10}$$

where NC is the number of characters in the result and DVP is the position within the result where the value of the dimension is shown. The Relevance Parameter Weight (RPW) incorporates the user's perception of relevance by rewarding the first attributes of the query $R_t(n(t))$ as highly desirable and penalising the last ones:

$$RPW = 1 - \frac{PD}{N} \tag{11}$$

where PD is the position of the n-attribute of the query $R_t(n(t))$ and N is the total number of dimensions of the query vector $R_t(n(t))$. The Dimension Parameter Weight (DPW) incorporates the observation of user relevance with the value of domains D(n(t)) by providing a better score on the domain values the user has more filled on the query:

$$DPW = \frac{NDT}{N} \tag{12}$$

where NDT is the number of dimensions with the same domain (word, number or price) on the query $R_t(n(t))$ and N is the total number of dimensions of the query vector $R_t(n(t))$. We assign this final Result Score value (RS) to each M-tuple v_o of the answer set A. This value is used by our ISA to reorder the answer set A of Y M-tuples, showing to the user the first set of Z results which have the higher potential value.

3.3   User Iteration

The user, based on the answer set A can now act as an intelligent critic and select a subset of P relevant results, C_P, of A. C_P is a set that consists of P M-tuples $C_P = \{v_1, v_2 \dots v_P\}$. We consider v_P as a vector of M dimensions; $v_p = (l_{p1}, l_{p2} \dots l_{pM})$ where l_p are the M different attributes for p = 1, 2..P. Similarly, the user can also select a subset of Q irrelevant results, C_Q of A, $C_Q = \{v_1, v_2 \dots v_Q\}$. We consider v_q as a vector of M dimensions; $v_q = (l_{q1}, l_{q2} \dots l_{qM})$ where lq are the M different attributes for q = 1, 2..Q. Based on the user iteration, our Intelligent Search Assistant provides to the user with a different answer set A of Z M-tuples reordered to MD, the minimum distance to the Relevant Centre for the results selected, following the formula:

$$RCP[n] = \frac{\sum_{p=1}^{P} SD_p[n]}{P} = \frac{\sum_{p=1}^{P} l_{pn}}{P} \tag{13}$$

where P is the number of relevant results selected, n the attribute index from the query $R_t(n(t))$ and $SD_p[n]$ the associated Score Dimension vector to the result or M-tuple v_P formed of l_{pn} attributes. An equivalent equation applies to the calculation of the Irrelevant Centre Point. Our Intelligent Search Assistant reorders the retrieved Y set of M-tuples showing only to the user the first Z set of M-tuples based on the lowest distance (MD) between the difference of their distances to both Relevant Centre Point (RD) and the Irrelevant Centre Point (ID) respectively:

$$MD = RD - ID \tag{14}$$

where MD is the result distance, RD is the Relevant Distance and ID is the Irrelevant Distance. The Relevant Distance (RD) of each result or M-tuple v_q is formulated as below:

$$RD = \sqrt{\sum_{n=1}^{N} (SD[n] - RCP[n])^2} \tag{15}$$

where SD[n] is the Score Dimension vector of the result or M-tuple v_q and RCP[n] is the coordinate of the Relevant Centre Point. Equivalent equation applies to the calculation of the Irrelevant Distance. Therefore we are presenting an iterative search progress that learns and adapts to the perceived user relevance based on the dimensions or attributes the user has introduced on the initial query.

3.4   Dimension Learning

The answer set A to the query $R_1(n(1))$ is based on the N dimension query introduced by the user however results are formed of M dimensions therefore the subset of results the user has considered as relevant may have other relevant concepts hidden the user did not considered on the original query. We consider the domain D(m) or the M attributes from which our universe U is formed as the different independent words that form the set of Y results retrieved from the search engines. Our cost function is expanded from the N attributes defined in the query $R_1(n(1))$ to the M attributes that form the searched results. Our Score Dimension vector, SD[m], is now based on M-dimensions. An analogue attribute expansion is applied to the Relevance Centre Calculation, RCP[m]. The query $R_1(n(1))$ is based on the N-Dimension vector introduced by the user however the answer set A consist of Y M-tuples. The user, based on the presented set A, selects a subset of P relevant results, C_P and a subset of Q irrelevant results, C_Q.

Lets consider C_P as a set that consists of P M-tuples $C_P = \{v_1, v_2 \ldots v_P\}$ where v_P is a vector of M dimensions; $v_P = (l_{p1}, l_{p2} \ldots l_{pM})$ and l_p are the M different attributes for $p = 1, 2..P$. The M-dimension vector Dimension Average, DA[m], is the average value of the m-attributes for the selected relevant P results:

$$DA[m] = \frac{\sum\limits_{p=1}^{P} SD_p[m]}{P} = \frac{\sum\limits_{p=1}^{P} l_{pm}}{P} \qquad (16)$$

where P is the number of relevant results selected, m the attribute index of the relation U and $SD_p[m]$ the associated Score Dimension vector to the result or M-tuple v_P formed of l_{pm} attributes. We define ADV as the Average Dimension Value of the M-dimension vector DA[m]:

$$ADV = \frac{\sum\limits_{m=1}^{M} DA[m]}{M} \qquad (17)$$

where M is the total number of attributes that form the relation U. The correlation vector σ[m] is the difference between the dimension values of each result with the average vector:

$$\sigma[m] = \frac{\sum\limits_{p=1}^{P} (SD_p[m] - DA[m])}{P} = \frac{\sum\limits_{p=1}^{P} (l_{pm} - DA[m])}{P} \qquad (18)$$

where P is the number of relevant results selected, m the attribute index of the relation U and $SD_p[m]$ the associated Score Dimension vector to the result or M-tuple v_P formed of l_{pm} attributes. We define C as the average correlation value of the M-dimensions of the vector σ[m]:

$$C = \frac{\sum_{m=1}^{M} \sigma[m]}{M} \tag{19}$$

where M is the total number of attributes that form the relation U. We consider an m-attribute relevant if its associated Dimension Average value DA[m] is larger than the average dimension ADV and its correlation value $\sigma[m]$ is lesser than the average correlation C. We have therefore changed the relevant attributes of the searched entities or ideas by correlating the error value of its concepts or properties represented as attributes or dimensions. On the next iteration, the query $R_2(n(2))$ is formed by the attributes our ISA has considered relevant. The answer to the query $R_2(n(2))$ is a different set A of Y M-tuples. This process iterates until there are not new relevant results to be shown to the user.

3.5 Gradient Descent Learning

Gradient Descent learning is based on the adaptation to the perceived user interests or understanding of meaning by correlating the attribute values of each result to extract similar meanings and cancel superfluous ones. The ISA Gradient Descent learning algorithm is based on a recurrent model. The inputs $i = \{i_1,...,i_P\}$ are the M-tuples v_P corresponding to the selected relevant result subset C_P and the desired outputs $y = \{y_1, ...,y_P\}$ are the same values as the input. Our ISA then obtains the learned random neural network weights, calculates the relevant dimensions and finally reorders the results according to the minimum distance to the new Relevant Centre Point focused on the relevant dimensions.

3.6 Reinforcement Learning

The external interaction with the environment is provided when the user selects the relevant result set C_P. Reinforcement Learning adapts to the perceived user relevance by incrementing the value of relevant dimensions and reducing it for the irrelevant ones. Reinforcement Learning modifies the values of the m attributes of the results, accentuating hidden relevant meanings and lowering irrelevant properties. We associate the Random Neural Network weights to the answer set A; W = A. Our ISA updates the network weights W by rewarding the result relevant attributes by:

$$w(p,m) = l_{pm}^{s-1} + l_{pm}^{s-1} * \left(\frac{l_{pm}^{s-1}}{\sum_{m=1}^{M} l_{pm}^{s-1}} \right) \tag{20}$$

where p is the result or M-tuple v_P formed of l_{pm} attributes, m the result attribute index, M the total number of attributes and s the iteration number. ISA also updates the network weights by punishing the result irrelevant attributes by:

$$w(p,m) = l_{pm}^{s-1} - l_{pm}^{s-1} * \left(\frac{l_{pm}^{s-1}}{\sum_{m=1}^{M} l_{pm}^{s-1}} \right) \tag{21}$$

where p is the result or M-tuple v_P formed of l_{pm} attributes, m the result attribute index, M the total number of attributes and s the iteration number. Our ISA then recalculates the potential of each of the result based on the updated network weights and reorders them, showing to the user the results which have a higher potential or score.

4   Validation

The Intelligent Internet Search Assistant we have proposed emulates how Web search engines work by using a very similar interface to introduce and display information. We validate our ISA algorithm with a set of three different experiments. Users in the experiments can both choose between the different Web search engines and the N number of results they would to retrieve from each one. We propose the following formula to measure Web search quality; it is based on the concept that a better search engine provides with a list of more relevant results on top positions. In an list of N results, we score N to the first result and 1 to the last result, the value of the quality proposed is then the summation of the position score based of each of the selected results. Our definition of Quality, Q, can be defined as:

$$Q = \sum_{i=1}^{Y} RSE_i \tag{22}$$

where RSE_i is the rank of the result i in a particular search engine with a value of N if the result is in the first position and 1 if the result is the last one. Y is the total number of results selected by the user. The best Web search engine would have the largest Quality value. We define normalized quality, $\overline{Q}$, as the division of the quality, Q, by the optimum figure which it is when the user consider relevant all the results provided by the Web search engine. On this situation Y and N have the same value:

$$\overline{Q} = \frac{Q}{\frac{N(N+1)}{2}} \tag{23}$$

We define I as the quality improvement between a Web search engine and a reference:

$$I = \frac{QW - QR}{QR} \tag{24}$$

where I is the Improvement, QW is the quality of the Web search engine and QR is the quality reference; we use the Quality of Google as QR in our validation exercise.

4.1   ISA Web Search Engine

In our first experiment, validators can select from which Web search engine they would their results to be retrieved from; as in our first experiment, the users need to select the relevant results. Our ISA combines the results retrieved from the different Web search engines selected. We present the average values for the 18 different queries. We show the normalized quality of each Web search engine selected including our ISA; because users can choose any Web search engine; we are not introducing the improvement value as we do not have a unique reference Web search engine (Table 1).

Table 1. Web search engine validation

Experiment 1–18 queries						
Web	Google	Yahoo	Ask	Lycos	Bing	ISA
Q̄	0.2691	0.2587	0.3454	0.3533	0.3429	0.4448

where Web term represents the Web Search Engines selected by the user and Q is the average Quality for the 18 different queries for each Web Search Engine including our ISA.

4.2   ISA Relevant Center Point

In our second experiment we have asked to our validators to search for different queries using only Google; ISA provides with a set of reordered results from which the user needs to select the relevant results. We show the average values for the 20 different queries, the average number of results retrieved by Google and the average number of results selected by the user. We represent the normalized quality of Google and ISA with the improvement of our algorithm against Google. In our third experiment, ISA provides with a reordered list from where the user needs to select which results are relevant. Our ISA reorders the results using the dimension relevant centre point providing to the user with another reordered result list from where the user needs to select the relevant ones. We show the average values for the 16 different queries, the average number of results selected by the user and the average number of results selected. We also represent the normalized quality of Google, ISA and the ISA with the relevant circle iteration including the improvement against Google in both scenarios (Table 2).

Table 2. Relevant center point validation

Experiment 2–20 queries						
Results retrieved	Results selected	Google Q	ISA Q	ISA I	ISA Circle Q	ISA Circle I
19.35	8.05	0.4626	0.4878	15.39 %	–	–
Experiment 3–16 queries						
Results retrieved	Results selected	Google Q	ISA Q	ISA I	ISA Circle Q	ISA Circle I
21.75	8.75	0.4451	0.4595	18 %	0.4953	26 %

where Experiment 2 and 3 results retrieved are the average results shown to the user, results selected are the average results the user considers relevant. Google and ISA Q are the average Quality values based on their different result list ranking. ISA I is the average improvement of our algorithm against Google. ISA Circle Q and I is the average Quality value with its associated Improvement after the first iteration where the user selects the relevant results and our algorithm reorder the results based on the minimum distance to the Relevant Centre Point.

4.3   ISA Learning

Users in this validation can choose between Google and Bing with either Gradient Descent or Reinforcement Learning type. Our ISA then collects the first 50 results from the Web search engine selected, reorders them according to its cost function and finally show to the user the first 20 results. We consider 50 results is a good approximation of search depth as more results can add clutter and irrelevance; 20 results is the average number of results read by a user before he launches another search if he does not find any relevant one.

Our ISA reorders results while learning on the two step iterative process showing only the best 20 results to the user. We present the average Quality values of the Web search engine and ISA for the 29 different queries searched by different users, the learning type and the Web search engine used (Table 3).

Table 3. Learning validation

Gradient descent learning: 17 queries								
First iteration			Second iteration			Third iteration		
Web	ISA	I	Web	ISA	I	Web	ISA	I
0.41	0.58	43 %	0.45	0.61	14 %	0.46	0.62	8 %
Reinforcement learning: 12 queries								
First iteration			Second iteration			Third iteration		
Web	ISA	I	Web	ISA	I	Web	ISA	I
0.42	0.57	34 %	0.47	0.67	36 %	0.49	0.68	0.0 %

where Web and ISA represent the Quality of the selected Web Search Engine and ISA respectively in the three successive learning iterations. The first I represents the improvement from ISA against the Web search; the second I is between ISA iterations 2 and 1 and finally the third I is between the ISA iterations 3 and 2.

5   Conclusions

We have defined a different process; the application of the Random Neural Network as a biological inspired algorithm to measure both user relevance and result ranking based on a predetermined cost function. We have proposed a novel approach to Web search where the user iteratively trains the neural network while looking for relevant results. Our Intelligent Search Assistant performs generally slightly better than Google and other Web search engines however, this evaluation may be biased because users tend to concentrate on the first results provided which were the ones we showed in our algorithm. Our ISA adapts and learns from user previous relevance measurements increasing significantly its quality and improvement within the first iteration. Reinforcement Learning algorithm performs better than Gradient Descent. Although Gradient Descent provides a better quality on the first iteration; Reinforcement Learning outperforms on the second one due its higher learning rate. Both of them have a residual learning on their third iteration. Gradient Descent would have been the preferred learning algorithm if only one iteration is required; however Reinforcement Learning would have been a better option in the case of two iterations. It is not recommended three iterations because learning is only residual. Deep learning may also be used [19]. Further work includes the validation of our Intelligent Search Assistant with more queries against other search engines such as metasearch engines, online academic databases and recommender systems. This validation comprises its ranking algorithm and its learning performance.

References

1. Wang, X., Zhang, L.: Search engine optimization based on algorithm of BP neural networks. In: Proceedings of the Seventh International Conference on Computational Intelligence and Security, pp. 390–394 (2011)
2. Boyan, J., Freitag, D., Joachims, T.: A machine learning architecture for optimizing Web search engines. In: Proceedings of the AAAI Workshop on Internet-Based Information Systems (1996)
3. Shu, B., Kak, S.: A neural network-based intelligent metasearch engine. Inf. Sci. Inform. Comput. Sci. **120**, 1–11 (2009)
4. Burgues, C., Shaked, T., Renshaw, E., Lazier, L., Deeds, M., Hamilton, N., Hullender, G.: Learning to rank using gradient descent. In: Proceedings of the 22nd International Conference on Machine Learning, ICML 2005, pp. 89–96 (2005)

5. Bermejo, S., Dalmau, J.: Web metasearch using unsupervised neural networks. In: Proceedings of the 7th International Work-Conference on Artificial and Natural Neural Networks: Part II: Artificial Neural Nets Problem Solving Methods, IWANN 2003, pp. 711–718 (2003)
6. Scarselli, F., Liang, S., Hagenbuchner, M., Chung, A.: Adaptive page ranking with neural networks. In: Proceedings of Special Interest Tracks and Posters of the 14th International Conference on World Wide Web, WWW 2005, pp. 936–937 (2005)
7. Gelenbe, E.: Random neural network with negative and positive signals and product form solution. Neural Comput. 1, 502–510 (1989)
8. Gelenbe, E.: Learning in the recurrent Random Neural Network. Neural Comput. 5, 154–164 (1993)
9. Gelenbe, E., Timotheou, S.: Random neural networks with synchronized interactions. Neural Comput. 20(9), 2308–2324 (2008)
10. Gelenbe, E.: Steps toward self-aware networks. Commun. ACM 52(7), 66–75 (2009)
11. Gelenbe, E., Wu, F.J.: Large scale simulation for human evacuation and rescue. Comput. Math Appl. 64(12), 3869–3880 (2012)
12. Filippoupolitis, A., Hey, L., Loukas, G., Gelenbe, E., Timotheou, S.: Emergency response simulation using wireless sensor networks. In: Proceedings of the 1st International Conference on Ambient Media and Systems, vol. 21 (2008)
13. Gelenbe, E., Koçak, T.: Area-based results for mine detection. IEEE Trans. Geosci. Remote Sens. 38(1), 12–24 (2000)
14. Cramer, C., Gelenbe, E., Bakircloglu, H.: Low bit-rate video compression with neural networks and temporal subsampling. Proc. IEEE 84(10), 1529–1543 (1996)
15. Atalay, V., Gelenbe, E., Yalabik, N.: The random neural network model for texture generation. Int. J. Pattern Recognit Artif Intell. 6(1), 131–141 (1992)
16. Gelenbe, E.: Search in unknown random environments. Phys. Rev. E 82, 061112 (2010)
17. Abdelrahman, O.H., Gelenbe, E.: Time and energy in team-based search. Phys. Rev. E 87 (3), 032125 (2013)
18. Gelenbe, E., Abdelrahman, O.H.: Search in the universe of big networks and data. IEEE Netw. 28(4), 20–25 (2014)
19. Gelenbe, E., Yin, Y.: Deep learning with random neural networks, paper number 16502. In: International Joint Conference on Neural Networks (IJCNN 2016), World Congress on Computational Intelligence, Vancouver, BC. IEEE Xplore (2016)

A Novel Grouping Genetic Algorithm for the One-Dimensional Bin Packing Problem on GPU

Sukru Ozer Ozcan, Tansel Dokeroglu$^{(\boxtimes)}$, Ahmet Cosar, and Adnan Yazici

Computer Engineering Department of Middle East Technical University,
Universities Street, 6800 Ankara, Turkey
{ozer.ozcan,tansel,cosar,yazici}@ceng.metu.edu.tr

Abstract. One-dimensional Bin Packing Problem (1D-BPP) is a challenging NP-Hard combinatorial problem which is used to pack finite number of items into minimum number of bins. Large problem instances of the 1D-BPP cannot be solved exactly due to the intractable nature of the problem. In this study, we propose an efficient Grouping Genetic Algorithm (GGA) by harnessing the power of the Graphics Processing Unit (GPU) using CUDA. The time consuming crossover and mutation processes of the GGA are executed on the GPU by increasing the evaluation times significantly. The obtained experimental results on 1,238 benchmark 1D-BPP instances show that our proposed algorithm has a high performance and is a scalable algorithm with its high speed fitness evaluation ability. Our proposed algorithm can be considered as one of the best performing algorithms with its 66 times faster computation speed that enables to explore the search space more effectively than any of its counterparts.

Keywords: 1D Bin packing · Grouping genetic · CUDA · GPU

1 Introduction

One-dimensional Bin Packing Problem (1D-BPP) is a challenging NP-Hard combinatorial problem which is used to pack finite number of items into minimum number of bins [1]. The general purpose of the 1D-BPP is to pack items of interest subject to various constraints such that the overall number of bins is minimized. More formally, 1D-BPP is the process of packing N items into bins which are unlimited in numbers and same in size and shape. The bins are assumed to have a capacity of $C > 0$, and items are assumed to have a size S_i for I in $\{1, 2, ..., N\}$ where $(S_i > 0)$. The goal is to find minimum number of bins in order to pack all of N items.

Although problems with a small number of items up to 30 can be solved with brute-force algorithms, large problem instances of the 1D-BPP cannot be solved exactly. Therefore, metaheuristic approaches such as genetic algorithms (GA), particle swarm, tabu search, and minimum bin slack (MBS) have been

© The Author(s) 2016
T. Czachórski et al. (Eds.): ISCIS 2016, CCIS 659, pp. 52–60, 2016.
DOI: 10.1007/978-3-319-47217-1_6

widely used to solve this important problem (near-) optimally [2–5]. Most of the state-of-the-art algorithms that have been proposed to solve the 1D-BPP are designed to run on a single processor and do not make use of the high performance computation opportunities that are offered by the recent parallel computation technologies. In this study, introduce an efficient Grouping Genetic Algorithm (GGA) by making use of the Graphics Processing Unit (GPU) using Compute Unified Device Architecture (CUDA) [6–9]. The population of solutions is kept on memory of GPU and the time consuming crossover, mutation, and fitness evaluation processes of the proposed GGA are also executed on the GPU. Therefore, a high performance heterogeneous computing environment is provided with a parallel computation support of GPU [10,11]. Our proposed algorithm is tested on 1,238 benchmark problem instances and has been observed to be a robust and scalable algorithm that can be considered as one of the best performing algorithms with its up to 66 times faster computation speed than the CPU-based version of GGAs. This talent of our proposed algorithm enables it to explore the solution space more effectively than any of its single-processor versions and obtain (near-)optimal results.

2 Proposed Algorithm (1D-BPP-CUDA)

Falkenaur's chromosome structure is chosen for our study due to its high performance [6,7].

Exon Shuffling Crossover: We use exon shuffling crossover [12], a recent technique borrowed from molecular genetics, for our proposed parallel algorithm. Molecular genetics is the field of biology and genetics that studies genes at a molecular level and employs methods to elucidate the molecular function and interactions among genes. An offspring is generated by a two phase crossover. In the first phase, all mutually exclusive segments are combined. In the second phase, the remaining items are used to build a new bin. During the execution of the algorithm, the exon shuffling crossover operations are run on the GPU.

The Mutation operator: enables new solutions using the current optimal solution. In this study, the mutation operator works based on the predefined mutation ratio. The number of groups chosen change depending on the population size and mutation ratio. The mutation operator works on a number of groups computed as multiplication of population size and the mutation ratio and select a number of groups randomly. The items of the selected groups are removed from the current solution list and they are added to remaining item list. At then end of mutation process, items in the remaining item list are inserted back to groups in the solution list using BFD algorithm.

Inversion operator: is applied to increase the transfer probability of fitter gene pair to the next generation. At the beginning of process, selected groups are interchanged [6]. The upcoming crossover and mutation operators take place on these interchanged sets. The inversion operator provides an increased opportunity for promising future generations without changing the item list during the operation.

Fitness function: gives us a value that is based on an equation defined by Falkenauer given below:

$$FF = \sum_{i=1}^{nb} \left(\frac{F_i}{c}\right)^k \tag{1}$$

There are different approaches to compute a fitness value in order to lead choice procedure. Some of the approaches to calculate fitness value increase the solution space by keeping suboptimal solutions. From the other side if we only prefer to use group size as the fitness value, better solutions can be discarded. As a result, the choice of fitness function (FF) requires additional caution. nb is the number of bins, F_i is the sum of weights of the elements packed into the bin i ($i = 1$,..., nb), c is the bin capacity, and k is a heuristic exponential factor. The value k expresses a concentration on the almost full bins in comparison to less filled ones. Falkenauer used $k = 2$ but Stawowy reported that $k = 4$ gives slightly better results therefore, we prefer the second value [15].

For calculating the fitness values of each chromosome, we prefer to have an enough block size division of size of population by 64 and 64 threads. So every chromosome's fitness value is calculated by concurrent blocks and threads. Communication between host and device has a price. Since item weights are constant values, it doesn't need to transfer back from device to host. But the population is needed to transfer from the device to host after the initial generation on the device for the truncating and adding BFD to the population. After these functions we need to transfer the population back again to the device to find slacks. For crossover, mutation and calculating fitness values, the population is transferred to the device again. Finally after the last function in the last generation on GPU, we transfer it back to host for validating and displaying the results. At that time we no longer need the Random Numbered Arrays, item values and population on the device. So, the final operation takes place on the device is to free the memory they are occupied on GPU.

For the mutation and generation of initial population, we need to generate integer random numbers. We use CURAND library of GPU side for this process. A basic generation of CURAND is used in our study. We send the state pointers to kernels to make the states ready for the generation-kernels. In this study, we use two different generation states to have completely different two 1000-element arrays. One of them generated by MTGP32 pseudorandom sequence generator which is an NVIDIA's adaption of an algorithm proposed by Saito et al. [13]. The other state we used is CURAND's default state which generates an array of pseudorandom numbers greater than 2^{190}. Kernel Concurrency and Host-Device Memory Copy Concurrency are used to do asynchronous operation for generation of two distinct random numbered arrays. Three streams are created totally in this step. First two of three are used for the generation, and the last one is used for asynchronous memory copy of item weights from host to device. These three operations are completely independent and run asynchronously.

An initial population is generated with the random numbered arrays for the proposed algorithm. After allocation enough memory on the device the kernel which executes the generation procedure, is launched. After the generation of each chromosome, the population array is filled with the chromosomes resident on the device memory.

Generating an initial population that is larger than the population that you will be working on by executing generations and pruning its size by selecting the fittest individuals is a very effective way for GA. With this method, it is possible to start with a higher quality population. This is called truncation. In our proposed algorithm, we applied this method on GPU. A number of random individuals are generated on the GPU and sent to CPU memory. CPU side code selected the best individuals by pruning the all initial population with a truncation ratio. The high-quality population is sent back to the GPU to be improved through the generations.

BFD is one of the simplest and high performance algorithms for solving the 1D-BPP. In our proposed algorithm, crossover and mutation operators use a BFD heuristic to reinsert the remaining items [4].

3   Performance Evaluation of Experimental Results

The PC used during the experiments has Intel Core i5-2467M CPU 1.60 GHz with 4 cores, 4 GB Memory (RAM), 64 bit Windows 7 Operating System, and EVGA NVIDIA GeForce GTX 750 Ti GPU (a mid-sized GPU designed for both gaming and computing environment).

Four different sets of problem instances are used during the experiments. The problem instances are set_1, set_2, set_3 [14] and hard28 [16] (Table 1).

Launching a kernel with N *Blocks* contains one *Thread* in each, equals to launching with one *Block* contains N *Thread* in terms of generating N software depended parallel processes. But execution times of each can be different for each configuration therefore, we set the best block and thread sizes to have a reasonable execution time.

The results of (near-)optimal population size for the Set_1 data set are presented in Table 2 (Bold face numbers are selected as the optimal solution, 80 individuals). *# of Optimal Solutions* shows the amount of optimal solution with comparing every instance with given optimal solutions for each data set instances. *Total Number of Extra Bins* shows the summation of extra bins which is calculated by subtracting found best solution, which is group/bin number required to pack all items, with the best solution for each data set instances. It is observed

Table 1. Information about the problem instances

problem instance	# instances	item weights	bin capacity (c)	# items (n)
set_1	720	[1,100]	{100, 120, 150}	{50, 100, 200, 500}
set_2	480	[3, 9] items at each bin	1,000	{50, 100, 200, 500}
set_3	10	[20,000, 35,000]	100,000	200
hard28	28	[1, 800]	1,000	{160, 180, 200}

Table 2. The effect of changing population size for Set_1 data set (# of generations is 40, truncate ratio is 20, mutation ratio is 0.2, inversion ratio is 0.2)

population size	# of optimal solutions	# of extra bins	execution time (sec.)
20	574	212	1239.00
40	584	174	1374.57
60	612	117	1570.78
80	**622**	**102**	**1696.64**
100	614	108	1701.86
150	613	110	2233.97
300	611	611	3712.35

that increase in population size has a limited effect on number of optimal solutions when number of generations is constant. The optimal number of population is selected for the remaining problem sets as it is performed on Set_1.

After finding the best population size for the algorithm, we performed tests on the number of generations to observe how it effects the solution quality and execution time of the algorithm. When we run the algorithm for this given set up on Set_1 data set, number of optimal solutions stays as 619 after the number of generation 40 and so the total number of extra bins required stays unchanged as expected. Additionally, execution time increases with the number of generations. The results for the Set_1 data set with each *Number of Generations* between 20 and 300 are presented in Table 3.

Mutation and inversion ratios correspond to the size of the array that will be generated in mutation and inversion processes. We tried to select the most effective ratios to find (near-)optimal solutions. The number of optimal solutions has an increasing pattern for Set_1 and Set_2 data sets. Additionally, an optimal number of solution 5 and extra number of bins 23 are found as a result for hard28 data set.

Table 3. The effect of changing the number of generations for Set_1 data set (# of population is 80, truncate ratio is 20, crossover ratio is 0.5, mutation ratio is 0.2, and inversion ratio is 0.2)

# of generations	# of optimal solutions	# of extra bins	execution time (sec.)
20	611	118	1038.01
40	**619**	**107**	**1282.00**
60	619	107	1457.57
80	619	107	1832.35
100	619	107	2205.55
150	619	107	3150.46
300	619	107	6171.10

Table 4. Comparisons between CPU and GPU implementation for Set_1 data set

population size	CPU-based exec.time	GPU-based exec.time	CPU solutions	GPU solutions	speed-up ratio
20	4852	773	547	571	6.28
40	5907	835	547	585	7.07
60	8296	927	547	610	8.95
80	10387	999	547	612	10.40
100	12897	1014	547	613	12.72

Table 5. Comparisons between CPU and GPU implementation for hard28 data set

population size	CPU-based exec.time	GPU-based exec.time	speed-up ratio
20	148	10.92	13.56
40	193	24.38	7.92
60	290	30.67	9.46
80	394	30.85	12.77
100	486	31.40	15.48
150	726	22.75	31.91
300	1434	21.58	**66.47**

The results of the comparisons made on problem Set_1 are presented in Table 4 for both CPU and GPU-parallel versions. Increasing the Population Size causes increase in the execution time for both CPU and GPU versions. The last column of Table 4 shows the Speed-Up Ratio. There is a constant increase in the *Speed Up Ratio*. For the data set_1, we have not only better solutions but have a speed up nearly 12 times approximately. In addition to that increase in the Population Size it does not have any effect on CPU implementation. The most important reason of this is to have a well distributed random generation of integers which provides us a wider search space of chromosomes and its groups.

Table 5 presents the speed-up performance of the proposed algorithm for the hard28 problem instances. The speed-up ratio is observed to be 66.47 for the problem set. The 1D-BPP-CUDA algorithm terminates the execution of the generations when it finds the optimal solution of the problem instance otherwise, it continues to search the solution space through larger number of generations. Therefore, the speed-up value of the algorithm is observed to be the highest on the problem set hard28 where obtained number of optimal solutions is less than the other problem sets and the number of generations are performed much more than the other problem sets.

As shown in the results, our algorithm both improves the solution quality while reducing the execution time even for a large population size and number of generations. In this section we compare our proposed algorithm with state-of-the-art algorithms in literature. Hard28 data set, one of the well known and widely

Table 6. Comparing the solution quality of GPU parallel 1D-BPP-GGA-CUDA algorithm with state-of-the-art algorithms on the hard28 data set

Algorithm	# of optimal solutions	Time (ms.)
BFD	2	2.3
MBS$'$	2	3.6
MBS	3	4.2
B2F	4	3.6
FFD	5	2.2
SAWMBS$'$	5	129.9
Pert-SAWMBS	5	6,946.4
Parallel Exon-MBS-BFD	5	5,341.0
1D-BPP-CUDA	5	**7,023.6**

preferred data set in BPP, is used for the comparisons [4]. See Table 6 for the results. This comparison may seem unfair however, we have parallel, sequential, GA and single solution versions of solutions in the same table. Yet, it may give a hint about execution times. A fair comparison can be made between Parallel Exon-MBS-BFD algorithm and our proposed 1D-BPP-CUDA algorithm.

With the (near-)optimal parameter settings of the 1D-BPP-GGA-CUDA algorithm, 84.57 % of the problem instances are solved optimally and the solutions found for each of the remaining problem instances produced only a single extra bin, which can be considered as high performance when compared with the sate-of-the-art algorithms.

4   Conclusions and Future Work

In this study, we propose a scalable heterogeneous computation based algorithm (1D-BPP-CUDA) that take advantage of CUDA, evolutionary grouping genetic metaheuristics, and bin-oriented heuristics to obtain high quality solutions for large scale 1D-BPP instances. A total number of 1,238 benchmark problems are examined with the proposed algorithm and it is shown that optimal solutions for 84.57 % of the problem instances can be obtained within practical optimization times while solving the rest of the problems with no more than one extra bin (250 additional bins in total). In addition to the higher solution quality, we have a speed-up of 66.47 times depending on the examined data set. When the results are compared with the existing state-of-the-art heuristics, the developed parallel hybrid grouping genetic algorithms can be considered among the best 1D-BPP algorithms in terms of computation time and solution quality.

References

1. Fleszar, K., Hindi, K.S.: New heuristics for one-dimensional bin-packing. Comput. Oper. Res. **29**(7), 821–839 (2002)
2. Cantu-Paz, E.: Efficient and Accurate Parallel Genetic Algorithms. Kluwer Academic Publishers, Dordrecht (2000)
3. Holland, J.H.: Adaptation in Natural and Artifical Systems. University of Michigan Press, Ann Arbor (1975)
4. Dokeroglu, T., Cosar, A.: Optimization of one-dimensional Bin Packing Problem with island parallel grouping genetic algorithms. Comput. Ind. Eng. **75**, 176–186 (2014)
5. Fernandez, A., Gil, C., Banos, R., Montoya, M.G.: A parallel multi-objective algorithm for two-dimensional bin packing with rotations and load balancing. Expert Syst. Appl. **40**(13), 5169–5180 (2013)
6. Falkenauer, E.: A new representation and operators for GAs applied to grouping problems. Evol. Comput. **2**(2), 123–144 (1994)
7. Falkenauer, E.: A hybrid grouping genetic algorithm for bin packing. J. Heurist. **2**(1), 5–30 (1996)
8. Quiroz-Castellanos, M., Cruz-Reyes, L., Torres-Jimenez, J., Gmez, C., Huacuja, H.J.F., Alvim, A.C.: A grouping genetic algorithm with controlled gene transmission for the bin packing problem. Comput. Oper. Res. **55**, 52–64 (2015)
9. Sivaraj, R., Ravichandran, T.: An efficient grouping genetic algorithm. Int. J. Comput. Appl. **21**(7), 38–42 (2011)
10. Zitzler, E., Thiele, L.: Multiobjective evolutionary algorithms: a comparative case study and the strength pareto approach. IEEE Trans. Evol. Comput. **3**(4), 257–271 (1999)
11. Dokeroglu, T.: Hybrid teaching-learning-based optimization algorithms for the Quadratic Assignment Problem. Comput. Ind. Eng. **85**, 86–101 (2015)
12. Rohlfshagen, P., Bullinaria, J.: A genetic algorithm with exon shuffling crossover for hard bin packing problems. In: Proceedings of the 9th Annual Conference on Genetic and Evolutionary Computation, pp. 1365–1371 (2007)
13. Saito, M., Matsumoto, M.: Variants of mersenne twister suitable for graphic processors. ACM Trans. Math. Softw. (TOMS) **39**(2), 12 (2013)
14. Scholl, A., Klein, R., Jurgens, C.: BISON: A fast hybrid procedure for exactly solving the one-dimensional bin packing problem. Comput. Oper. Res. **24**(7), 627–645 (1997)

15. Stawowy, A.: Evolutionary based heuristic for bin packing problem. Comput. Ind. Eng. **55**, 465–474 (2008)
16. Belov, G., Scheithauer, G., Mukhacheva, E.A.: One-dimensional heuristics adapted for two-dimensional rectangular strip packing. J. Oper. Res. Soc. **59**(6), 823–832 (2007)

Data Classification and Processing

A Novel Multi-criteria Inventory Classification Approach: Artificial Bee Colony Algorithm with VIKOR Method

Hedi Cherif[1(✉)] and Talel Ladhari[1,2]

[1] Ecole supérieure des Sciences Economiques et Commerciales de Tunis, Université de Tunis, Tunis, Tunisia
cherif.hedi@gmail.com
[2] College of Business, Umm Al-Qura University, Umm Al-Qura, Saudi Arabia

Abstract. ABC analysis is a well-established categorization technique based on the Pareto Principle which dispatches all the items into three predefined and ordered classes A, B and C, in order to derive the maximum benefit for the company. In this paper, we present a new approach for the ABC Multi-Criteria Inventory Classification problem based on the Artificial Bee Colony (ABC) algorithm with the Multi-Criteria Decision Making method namely VIKOR. The ABC algorithm tries to learn and optimize the criteria weights, which are then used as an input parameters by the method VIKOR. The MCDM method generates a ranking items and therefore an ABC classification. Each established classification is evaluated by an estimation function, which also represents the objective function. The results of our proposed approach were obtained from a widely used data set in the literature, and outperforms the existing classification models from the literature, by obtaining better inventory cost.

Keywords: ABC multi-criteria inventory classification · Hybrid model · Artificial Bee Colony · VIKOR

1 Introduction

The ABC analysis is a popular and widely used technique for the inventory classification problem and categorizes inventory items into three groups: A, B, or C based on some criteria in order to establish appropriate levels of control over each group.

For some time now, several metaheuristics have been deployed to tackle the MCIC problem. Tsai and Yeh [31] uses the particle swarm optimization technique and presents an inventory classification algorithm that simultaneously search the optimum number of inventory classes and perform classification, while Mohammaditabar et al. deploys the simulating annealing method [24] and proposes an integrated model to categorize the items and at the same time find the best policy. Saaty [30] has developed the Analytic Hierarchy Process (AHP) method,

© The Author(s) 2016
T. Czachórski et al. (Eds.): ISCIS 2016, CCIS 659, pp. 63–71, 2016.
DOI: 10.1007/978-3-319-47217-1_7

which has been widely deployed by some researchers to tackle the MCIC problem [4,7,8,27,28]. Other researchers [12,14,15] used a fuzzy version of the AHP method (FAHP). Lolli et al. [22] established a multi-criteria classification model called AHP-K-Veto, based on the AHP method and the K-means algorithm. Bhattacharya et al. [3] developed a model which combines Topsis (Technique for Order Preferences by Similarity to the Ideal Solution) with AHP method in order to generate a ranking items and then an ABC classification. Chen et al. [6] proposed an alternative approach to MCIC problem by using Topsis and two virtual items. Guvenir and Erel [9] developed a method that uses the generic algorithm for the sake of learning criteria weight, and established cut-off points between the classes A-B and B-C, in order to generate a classification of items. Then they showed in a second study [10] that their method based on genetic algorithm gives better performance than the AHP method. Al-Obeidat et al. [1] proposed an hybrid model which combine Differential Evolution method with the PROAFTN method, by using the evolutionary algorithm to inductively obtain PROAFTN's parameters from data to achieve a high classification accuracy. Liu et al. [21] combined the methods of Electre III and the simulating annealing to deal with the compensatory effect of the items against criteria and opted for grouping criteria. A new MCDM method of Evaluation based on Distance from Average Solution (called EDAS) [19] is introduced by calculating the best alternative according to the distance from positive and negative solutions.

To the best of our knowledge, the Artificial Bee Colony algorithm [16–18] and the VIKOR method [23,26,32] were not used to solve the ABC MCIC problem. In this paper, we present a new hybrid approach based on these two methods, which attempt to combine the main advantages of each used method. In our approach, the multi-criterion decision problem is modeled by using Vikor model whose parameters are tuned by using a bee colony optimization algorithm. Each established classification is evaluated by using an estimation function based on the inventory cost and the fill rate service level [2], which also represents the objective function of our model, by minimizing the classification cost.

The rest of the paper is organized as follows. In Sect. 2, the ABC algorithm and the VIKOR method are briefly presented. We also describe our proposed hybrid optimization model by adapting the ABC algorithm to be in compliance with the constraints of the problem. Section 3 presents the experimental results and a comparative numerical study with some models from the literature, based on a widely used dataset. We end this paper with a conclusion and discussion regarding future research.

2   The Proposed Work

2.1   Artifical Bee Colony Algorithm

The Artificial Bee Colony optimization algorithm which belongs to the family of evolutionary algorithms is based on a particular intelligent behavior of swarms bees. This approach is inspired by the real behaviors of the bees in their food research and how to share the information on the location of these food sources with other bees from the hive. The method classify the artificial bees into three distinct groups with specific tasks for each category of bees (employed bees, onlookers and scouts). More explicitly, the ABC algorithm is defined by the following steps:

- **Initialization:** We begin by generating randomly the initial population in the search space following this equation:

$$u_j = x_j^{min} + rand[0,1](x_j^{max} - x_j^{min})$$ (1)

Where x_j^{min} and x_j^{max} represent the bounds of the search space and $rand[0,1]$ generates a random number between 0 and 1. Each employed bee evaluate the nectar amount of a food source corresponding to the quality (Fitness) of the associated solution, by:

$$Fitness(i) = \frac{1}{F_{Objective}(i)}$$ (2)

Where $F_{Objective}$ represents the objective function used in our approch.

- **Moving onlooker bees:** The onlookers choose a food source according to the probability value associated with this food source, denoted P_i and calculated by the following expression:

$$P_i = \frac{Fitness(i)}{\sum_{n=1}^{SN} Fitness(n)}$$ (3)

where SN is the number of food sources equal to the number of employed bees. The onlooker bee selects a food source and then evaluates its amounts of nectar. Then, the bee moves according to the following formula:

$$v_{ij} = x_{ij} + \phi_{ij}(x_{ij} - x_{kj})$$ (4)

Where $k \in [1, 2, ..., SN]$ and $j \in [1, 2, ..., D]$ are randomly chosen indexes. Although k is determined randomly, it has to be different from i. ϕ_{ij} is a random number between $[-1, 1]$. It controls the production of neighbor food sources around x_{ij}.

- **Moving scout bees:** If the values of the fitness function of employed bees are not improved for a predetermined number of iterations ($Limit$), these food sources are abandoned, and the bee that is in this area will move randomly to explore other new food sites, hence the conversion of employed bee to the scout bees. The movement is done by this following equation:

$$v_{ij} = v_{ij}^{min} + rand[0,1](v_{ij}^{max} - v_{ij}^{min})$$ (5)

At each iteration, the solution having the best value of the Fitness function and the position of the food source found by bees are saved. All these steps are repeated for a predefined number of iterations or until a stopping criterion is satisfied.

2.2   VIKOR

Starting with criteria weights, the VIKOR method operates in order to obtain a compromise ranking-list, as well as the compromise solution, and determines the weight stability intervals for preference stability of the compromise solution. The basic idea of this MCDM method is that the ranking items is based on an index, computed from the measure of "closeness" to the "ideal" solution [23,26,32].

We consider that the value of the ith criterion function for the alternative a_j is denoted f_{ij} and the alternatives are denoted $a_1, a_2, ..., a_J$, with n as the total number of criteria. f_i^* and f_i^- represents the best and the worst values of all criterion functions, and B and C represent respectively the sets of benefit and cost criteria. The VIKOR method use the following $L_p - metric$ as an aggregate function:

$$L_{p,j} = \left\{ \sum_{i=1}^{n} [(w_j(f_i^* - f_{ij})/(f_i^* - f_i^-)]^p \right\}^{\frac{1}{p}} \quad 1 \le p \le \infty; j = 1, 2, ..., J. \quad (6)$$

$$f_i^* = \{(max_j\{f_{ij}\}|j \in B, min_j\{f_{ij}\}|j \in C)\} \quad (7)$$

$$f_i^- = \{(min_j\{f_{ij}\}|j \in B, max_j\{f_{ij}\}|j \in C)\} \quad (8)$$

The next step consists of calculating the three measures S, R and Q (VIKOR Index) of compromise ranking method VIKOR and sort all the alternatives according to these 3 ordered lists:

$$S_j = \sum_{i=1}^{n} w_j(f_i^* - f_{ij})/(f_i^* - f_i^-) \quad (9)$$

$$R_j = max_i \left[(w_j(f_i^* - f_{ij})/(f_i^* - f_i^-) \right] \quad (10)$$

$$Q_j = v(S_j - S^*)/(S^- - S^*) + (1 - v)(R_j - R^*)/(R^- - R^*)$$

$$S^* = min_j S_j, \qquad S^- = max_j S_j, \qquad R^* = min_j R_j, \qquad R^- = max_j R_j. \quad (11)$$

w_i are the criteria weights and v represents a factor used by the decision maker and reflects the weight of the strategy of "the maximum group utility". By convention, this factor v is set to 0.5. Once the VIKOR indexes Q_j, S_j and R_j are calculated, it only remains to sort all the alternatives in decreasing order of the values S, R and Q, for the purpose of obtaining three ranking lists. The VIKOR algorithm proposes as a compromise solution, for given criteria weights, the alternative (a'), which is the best ranked by measure Q, if a two conditions are satisfied [26].

2.3   A New Hybrid Approach for ABC MCIC

We present our proposed hybrid approach developed for the ABC MCIC problem. First, we describe the adjustments made to the Artificial Bee Colony algorithm, in order to comply with the constraints of the problem. The ABC algorithm initializes a population of solutions where each solution has D parameters. These parameters are generated respectively according to the Eqs. 1 and 5 and each vector represents a candidate solution for the optimization problem. But, given that the sum of these generated value may be different from 1, we used a generation procedure of initial solutions to adjust these values according to the constraints of VIKOR method, using the following equation:

$$x_{i,j} = x_{max} - rand\left[0, \left[x_{max} - \sum_{t=1}^{D} x_{i,t}\right]\right] \tag{12}$$

This formulation ensures whenever the sum of the solution parameters is equal to 1. When the onlooker bee move (Eq. 4), the mutation operation of ABC algorithm must be adapted, because the values of the generated solution can overflow the search space. To address this ambiguity, we calibrated the values so that the parameters are still within the range of our required search space:

$$X_{i,j} = \begin{cases} 0 & \text{if } X_{i,j} < 0 \\ 1 & \text{if } X_{i,j} > 1 \\ X_{i,j} & \text{otherwise.} \end{cases} \tag{13}$$

This adjustment values can still generate values that their sum is not equal to 1. In this sense, we proceeded to the normalization of the vector to achieve a unitary sum, using the following equation:

$$X_{i,j} = \frac{X_{i,j}}{\sum_{z=1}^{D} X_{i,z}} \tag{14}$$

Once these solutions are generated by the ABC algorithm, they will be considered by the VIKOR method as an input parameters, to calculate a score for each item, establish a total ranking items and consequently generate an ABC classification (according to the 20 %–30 %–50 % ABC distribution).

3   Experimental Results

To evaluate the performance of our proposed hybrid approach in the ABC MCIC problem, we consider a data set provided by an Hospital Respiratory Therapy Unit (HRTU). This data set has been widely used in the literature and contains 47 inventory items evaluated in terms of three criteria. This data set is displayed in the Table 1. The ABC classification results of the existing ABC classification models (R model [29], ZF model [33], Chen model [5], H model [11], NG model [25], ZF-NG model [20] and ZF-H model [13]) and our model are showed also in Table 1. Note that all the established classifications respect the same ABC distribtion, with 10 items in the class A, 14 items in the class B and 23 items in the class C. We clearly observe that our proposed approach provides a more efficient classification cost (833.677) than all other models presented from the literature, with a good Fill Rate (0.972) reflecting a good classification.

Table 1. Our approach vs existing classification models

Item	ADU	AUC	LT	R [29]	ZF [33]	Chen [5]	H [11]	NG [25]	ZF-NG [20]	ZF-H [13]	ABC-Vikor
1	5840.64	49.92	2	A	A	A	A	A	A	C	**C**
2	5670	210	5	A	A	A	A	A	A	A	**A**
3	5037.12	23.76	4	A	A	A	A	A	A	A	**C**
4	4769.56	27.73	1	B	C	B	A	A	B	B	**C**
5	3478.8	57.98	3	B	B	B	A	A	A	A	**C**
6	2936.67	31.24	3	C	C	B	B	A	B	B	**C**
7	2820	28.2	3	C	C	B	B	B	B	B	**C**
8	2640	55	4	B	B	B	B	B	B	A	**B**
9	2423.52	73.44	6	A	A	A	A	A	A	A	**A**
10	2407.5	160.5	4	B	A	A	A	A	A	A	**A**
11	1075.2	5.12	2	C	C	C	C	C	C	A	**C**
12	1043.5	20.87	5	B	B	B	B	B	B	B	**C**
13	1038	86.5	7	A	A	A	A	A	A	A	**A**
14	883.2	110.4	5	B	A	B	A	B	A	A	**A**
15	854.4	71.2	3	C	C	C	C	C	C	B	**B**
16	810	45	3	C	C	C	C	C	C	B	**C**
17	703.68	14.66	4	C	C	C	C	C	C	C	**C**
18	594	49.5	6	A	A	B	B	B	B	B	**B**
19	570	47.5	5	B	B	B	B	B	B	B	**B**
20	467.6	58.45	4	C	B	C	C	C	C	B	**B**
21	463.6	24.4	4	C	C	C	C	C	C	C	**C**
22	455	65	4	C	B	C	C	C	C	B	**B**
23	432.5	86.5	4	C	B	C	B	B	B	B	**A**
24	398.4	33.2	3	C	C	C	C	C	C	C	**C**
25	370.5	37.05	1	C	C	C	C	C	C	C	**C**
26	338.4	33.84	3	C	C	C	C	C	C	C	**C**
27	336.12	84.03	1	C	C	C	C	C	C	C	**C**
28	313.6	78.4	6	A	A	A	B	B	A	B	**A**
29	268.68	134.34	7	A	A	A	A	A	A	A	**A**
30	224	56	1	C	C	C	C	C	C	C	**C**
31	216	72	5	B	B	B	B	B	B	B	**A**
32	212.08	53.02	2	C	C	C	C	C	C	C	**C**
33	197.92	49.48	5	B	B	B	B	B	B	C	**B**
34	190.89	7.07	7	A	B	A	B	B	B	C	**B**
35	181.8	60.6	3	C	C	C	C	C	C	C	**B**
36	163.28	40.82	3	C	C	C	C	C	C	C	**C**
37	150	30	5	B	B	B	C	C	C	C	**B**
38	134.8	67.4	3	C	C	C	C	C	C	C	**B**
39	119.2	59.6	5	B	B	B	B	B	B	C	**B**
40	103.36	51.68	6	B	B	B	B	B	B	C	**A**
41	79.2	19.8	2	C	C	C	C	C	C	C	**C**
42	75.4	37.7	2	C	C	C	C	C	C	C	**C**
43	59.78	29.89	5	B	C	C	C	C	C	C	**B**
44	48.3	48.3	3	C	C	C	C	C	C	C	**C**
45	34.4	34.4	7	A	B	A	B	B	B	B	**B**
46	28.8	28.8	3	C	C	C	C	C	C	C	**C**
47	25.38	8.46	5	B	C	C	C	C	C	C	**C**
Classification cost				927.517	945.357	958.143	999.892	1011.007	985.599	971.018	**833.677**
Fill Rate				0.986	0.984	0.988	0.99	0.991	0.989	0.989	**0.972**

4 Conclusions

In this paper, we present a new hybrid approach for ABC MCIC problem. The main contribution of the proposed work is to exploit the efficiency of the ABC algorithm and the method VIKOR on a hybrid manner, to classify the inventory items based on objective weights and to reduce the inventory cost. A comparison has been made between the proposed approach and some existing methods and showed the good performance of the proposed method that outperforms some models from the literature. The idea of combining these two methods in our approach can be easily applied to general multi-criteria classification problems, not just the ABC MCIC problem. To extend this research, it would be interesting to assess the benefits of applying our model empirically using larger datasets.

References

1. Al-Obeidat, F., Belacel, N., Carretero, J.A., Mahanti, P.: Differential evolution for learning the classification method proaftn. Knowl.-Based Syst. **23**(5), 418–426 (2010)
2. Babai, M., Ladhari, T., Lajili, I.: On the inventory performance of multi-criteria classification methods: empirical investigation. Int. J. Prod. Res. **53**(1), 279–290 (2015)
3. Bhattacharya, A., Sarkar, B., Mukherjee, S.: Distance-based consensus method for ABC analysis. Int. J. Prod. Res. **45**(15), 3405–3420 (2007)
4. Braglia, M., Grassi, A., Montanari, R.: Multi-attribute classification method for spare parts inventory management. J. Qual. Mainten. Eng. **10**(1), 55–65 (2004)
5. Chen, J.: Peer-estimation for multiple criteria ABC inventory classification. Comput. Oper. Res. **38**(12), 1784–1791 (2011)
6. Chen, J.X.: Multiple criteria ABC inventory classification using two virtual items. Int. J. Prod. Res. **50**(6), 1702–1713 (2012)
7. Flores, B., Olson, D., Dorai, V.: Management of multicriteria inventory classification. Math. Comput. Model. **16**(12), 71–82 (1992)
8. Gajpal, P., Ganesh, L., Rajendran, C.: Criticality analysis of spare parts using the analytic hierarchy process. Int. J. Prod. Econ. **35**(1), 293–297 (1994)
9. Güvenir, H.A.: A genetic algorithm for multicriteria inventory classification. In: Artificial Neural Nets and Genetic Algorithms, pp. 6–9. Springer, Vienna (1995)

10. Guvenir, H.A., Erel, E.: Multicriteria inventory classification using a genetic algorithm. Eur. J. Oper. Res. **105**(1), 29–37 (1998)
11. Hadi-Vencheh, A.: An improvement to multiple criteria abc inventory classification. Eur. J. Oper. Res. **201**(3), 962–965 (2010)
12. Hadi-Vencheh, A., Mohamadghasemi, A.: A fuzzy ahp-dea approach for multiple criteria abc inventory classification. Expert Syst. Appl. **38**(4), 3346–3352 (2011)
13. Kaabi, H., Jabeur, K.: A new hybrid weighted optimization model for multi criteria ABC inventory classification. In: Abraham, A., Wegrzyn-Wolska, K., Hassanien, A.E., Snasel, V., Alimi, A.M. (eds.) AECIA 2015. AISC, vol. 427, pp. 261–270. Springer, Heidelberg (2016). doi:10.1007/978-3-319-29504-6_26
14. Kabir, G.: Multiple criteria inventory classification under fuzzy environment. Int. J. Fuzzy Syst. Appl. (IJFSA) **2**(4), 76–92 (2012)
15. Kabir, G., Hasin, M.: Multiple criteria inventory classification using fuzzy analytic hierarchy process. Int. J. Ind. Eng. Comput. **3**(2), 123–132 (2012)
16. Karaboga, D., Akay, B.: A comparative study of artificial bee colony algorithm. Appl. Math. Comput. **214**(1), 108–132 (2009)
17. Karaboga, D., Basturk, B.: On the performance of artificial bee colony (ABC) algorithm. Appl. Soft Comput. **8**(1), 687–697 (2008)
18. Karaboga, D., Gorkemli, B., Ozturk, C., Karaboga, N.: A comprehensive survey: artificial bee colony (abc) algorithm and applications. Artif. Intell. Rev. **42**(1), 21–57 (2014)
19. Ghorabaee Keshavarz, M., Zavadskas, E.K., Olfat, L., Turskis, Z.: Multi-criteria inventory classification using a new method of evaluation based on distance from average solution (edas). Informatica **26**(3), 435–451 (2015)
20. Ladhari, T., Babai, M., Lajili, I.: Multi-criteria inventory classification: new consensual procedures. IMA J. Manag. Math., dpv003 (2015)
21. Liu, J., Liao, X., Zhao, W., Yang, N.: A classification approach based on the outranking model for multiple criteria abc analysis. Omega (2015)
22. Lolli, F., Ishizaka, A., Gamberini, R.: New AHP-based approaches for multi-criteria inventory classification. Int. J. Prod. Econ. **156**, 62–74 (2014)
23. Mardani, A., Zavadskas, E.K., Govindan, K., Amat Senin, A., Jusoh, A.: Vikor technique: a systematic review of the state of the art literature on methodologies and applications. Sustainability **8**(1), 37 (2016)
24. Mohammaditabar, D., Ghodsypour, S., O'Brien, C.: Inventory control system design by integrating inventory classification and policy selection. Int. J. Prod. Econ. **140**(2), 655–659 (2012)
25. Ng, W.L.: A simple classifier for multiple criteria abc analysis. Eur. J. Oper. Res. **177**(1), 344–353 (2007)
26. Opricovic, S.: Multicriteria optimization of civil engineering systems. Fac. Civil Eng. Belgrade **2**(1), 5–21 (1998)
27. Partovi, F., Burton, J.: Using the analytic hierarchy process for ABC analysis. Int. J. Oper. Prod. Manag. **13**(9), 29–44 (1993)
28. Partovi, F., Hopton, W.: The analytic hierarchy process as applied to two types of inventory problems. Prod. Invent. Manag. J. **35**(1), 13 (1994)
29. Ramanathan, R.: Abc inventory classification with multiple-criteria using weighted linear optimization. Comput. Oper. Res. **33**(3), 695–700 (2006)
30. Saaty, T.: The Analytical Hierarchy Process: Planning, Setting Priorities, Resource Allocation. McGraw-Hill, New York (1980)

31. Tsai, C.Y., Yeh, S.W.: A multiple objective particle swarm optimization approach for inventory classification. Int. J. Prod. Econ. **114**(2), 656–666 (2008)
32. Yazdani, M., Graeml, F.R.: Vikor and its applications: a state-of-the-art survey. Int. J. Strat. Dec. Sci. (IJSDS) **5**(2), 56–83 (2014)
33. Zhou, P., Fan, L.: A note on multi-criteria abc inventory classification using weighted linear optimization. Eur. J. Oper. Res. **182**(3), 1488–1491 (2007)

Android Malware Classification by Applying Online Machine Learning

Abdurrahman Pektaş[1], Mahmut Çavdar[2], and Tankut Acarman[2(✉)]

[1] The Scientific and Technological Research Council of Turkey, Ankara, Turkey
[2] Computer Engineering Department, Galatasaray University, İstanbul, Turkey
acarmant@gmail.com

Abstract. A malware is deployed to execute malicious activities in the compromised operating systems. The widespread use of android smartphones with high speed Internet and permissions granted to applications for accessing internal logs provides a favorable environment for the execution of unauthorized and malicious activities. The major risk and challenge lies along classification of a large volume and variety of malware. A malware may evolve and continue to hide its malicious activies against security systems. Knowing malware features a priori and classification of a malware plays a crucial role at defending the safety and liability critical user's information. In this paper, we study android malware activities, features and apply online machine learning algorithm to classify a new android malware. We extract a fairly adequate set of malware features and we evaluate a machine learning based classification method. The runtime model is built and it can be implemented to detect variants of an android malware. The metrics illustrate the effectiveness of the proposed classification method.

1 Introduction

According to Internet Security Report, 1.4 billion smartphones were sold in 2015 and 83,3 % phones were running Android, [1]. Their users may save information about their personal identities, online payment system access and user's credentials. Malware authors, cyber criminals aim to steal these information via the distribution and installation of android applications. Overall, 3.3 million applications were classified as malware in 2015. Malware authors deliver this large variety and volume of malicious software by using advanced obfuscation techniques. Therefore, behavior-based malware analysis and classification of a malware sample to its original family plays a crucial and timely role at taking security and protection counter measures.

Android is a complete operating system that uses Android application (app) package (APK) for distribution and installation of mobile apps. APK file contains components which share a set of resources like database, preference, files, classes compiled in the dex file format, etc., App components are divided in four categories: *activities* handling the user interaction; *services* carrying out background tasks; *content providers* managing app's data; *broadcast receivers*

T. Czachórski et al. (Eds.): ISCIS 2016, CCIS 659, pp. 72–80, 2016.
DOI: 10.1007/978-3-319-47217-1_8

Table 1. List of system commands and command's execution frequency by our malware test set

Command	Description	Frequency
/system/bin/cat (i.e. cat)	display files	33
logCat	reads the compressed logging files and outputs human-readable messages	13
ping	verifies IP-level connectivity by using ICMP	6
chmod	used to change the permissions of files or directories	4
ln	creates a link to an existing file	3
mount	attaches additional filesystem	2
echo	outputs text to the screen or a file	2
su	used to execute commands with the privileges of another account	2
id	print user ID and group ID of the current user	2

assuring communications between components, app's, even more Android OS. The manifest declares the app's components and how they interact. Also user permissions required by the apps are placed in the manifest file. Android is a privilege-separated operating system, in which each application runs with a distinct system identity (Linux user ID and group ID). Parts of the system are also separated into distinct identities. Linux thereby isolates applications from each other and from the system.

Several commands can be used to infect Android devices. For example, Cat command, i.e., System/bin/cat displays files in the system and it can be executed for malicious purposes. The command-line tool LogCat can be used for viewing the internal logs. Log messages may include privacy-related information. An app can access the log file by giving every app the READ_LOGS permission with aid of the *chmod* command. The list of commands is described in Table 1.

In line with the emerging market of android smartphones, detection and classification of its malware has attracted a lot of attention. Static analysis of the executables by using commands, and modelling of malware features by using permissions and API calls is presented for the detection of a malware in [2,3]. K-means algorithm for clustering and a decision tree learning algorithm for classification of a malware is presented by monitoring various permission based features and events extracted from applications in [4]. A learning model database is obtained by collecting the extracted features and N-gram signatures are created in [5]. Text mining and information retrieval is applied for the static analysis of a malware in [6]. In [7], a heuristics approach by using 39 different behaviour flags such as Java API calls, presence of embedded executables and code size is developed to determine whether an application is malicious or not. A deep learning for automatic generation of malware signature is studied to detect a majority of new variants of a malware in [8]. And, a detection model is trained with the information gathered via the communication among components. A security framework has been deployed by an European project called NEMESYS for gathering and analyzing information about the nature of cyber-attacks targeting mobile devices and presented a model-based approach for detection of anomalies [9–11].

The paper is organized as follows: In Sect. 2, we present the selected features. In Sect. 3, we implement online machine learning algorithm to the classification of malware samples and we evaluate the results. Finally, we conclude our paper.

2   Feature Set

Cuckoo Sandbox is an open source analysis system and relies on virtualization technology to run a given file, [12]. It can analyze both executable and non-executable files and monitor the run-time activities. In this study, we extracted

Table 2. Features and their types

Feature category	Type	Value
commands	String	/system/bin/cat
services	String	com.houseads.AdService, com.applovin.sdk.AppLovinService,'
fingerprint	String	getSimCountryIso, getDeviceId, getLine1Number
permissions	String	INTERNET, ACCESS_NETWORK_STATE, READ_PHONE_STATE, GET_ACCOUNTS
data_leak	String	getAccounts
file_accessed	String	/proc/net/if_inet6, /proc/meminfo ...
httpConnections	String	http://houseads.eu/ads/new_user.php?id=147 &im= 351451208401216 &l=en&c=us&bm =Nexus+5&bv=4.1.2&v=4.2&ct=UMTS &a=null&ts=04032016070451&m=&s=16
send_sms	Boolean	FALSE
receive_sms	Boolean	FALSE
read_sms	Boolean	FALSE
call_phone	Boolean	FALSE
ap_execute_shell_commands	Boolean	TRUE
app_queried_account_info	Boolean	TRUE
app_queried_installed_apps	Boolean	FALSE
app_queried_phone_number	Boolean	TRUE
app_queried_private_info	Boolean	FALSE
app_recording_audio	Boolean	FALSE
app_registered_receiver_runtime	Boolean	TRUE
app_uses_location	Boolean	FALSE
embedded_apk	Boolean	FALSE
is_dynamic_code	Boolean	TRUE
is_native_code	Boolean	FALSE
is_reflection_code	Boolean	TRUE

Table 3. Top 20 requested permissions

Permissions	Frequency
INTERNET	867
READ_PHONE_STATE	826
WRITE_EXTERNAL_STORAGE	764
ACCESS_NETWORK_STATE	744
SEND_SMS	565
INSTALL_SHORTCUT	535
ACCESS_WIFI_STATE	524
WAKE_LOCK	473
RECEIVE_BOOT_COMPLETED	420
VIBRATE	382
RECEIVE_SMS	348
GET_TASKS	337
WRITE_SETTINGS	306
READ_SMS	285
ACCESS_COARSE_LOCATION	281
READ_SETTINGS	278
CHANGE_WIFI_STATE	277
ACCESS_FINE_LOCATION	270
CALL_PHONE	215
SYSTEM_ALERT_WINDOW	182

the most significant and distinguishing behavioral features from the Cuckoo's analysis report. The list of android malware features is given in Table 2. The permissions requested by the applications are ranked according to their persistency in Table 3.

3   Implementation

The testing malware dataset is obtained from "VirusShare Malware Sharing Platform" ([13]), which provides a huge amount of different type malware including PE, HTML, Flash, Java, PDF, APK etc. All experiments were conducted under the Ubuntu 14.04 Desktop operating system with Intel(R) Core(TM) i5-2410M@2.30 GHz processor and 2 GB of RAM. The analysis with 5 guest machines took 5 days to analyze approximately 2000 samples. For labeling malware samples, we used Virustotal, an online web-based multi anti-virus scanner, [14]. The malware classes along their class-specific measures are given in Table 4.

Table 4. Malware families and their class-specific measures

Family	Code	#	Recall	Specificity	Precision	Balanced accuracy
android.trojan.fakeinst	1	193	0.94	0.98	0.94	0.96
android.riskware.smsreg	2	104	0.67	0.99	0.86	0.83
android.trojan.agent	3	79	0.60	1.00	1.00	0.80
android.adware.gingermaster	4	74	0.67	0.99	0.80^d	0.83
android.adware.adwo	5	69	0.83	1.00	1.00	0.92
android.trojan.smssend	6	66	1.00	0.84	0.35	0.92
android.trojan.smskey	7	48	0.25	1.00	1.00	0.63
android.adware.utchi	8	45	1.00	1.00	1.00	1.00
android.trojan.clicker	9	37	1.00	0.99	0.75	0.99
android.adware.appquanta	10	34	1.00	1.00	1.00	1.00
android.adware.plankton	11	34	0.50	1.00	1.00	0.75
android.trojan.fakeapp	12	19	1.00	1.00	1.00	1.00
android.trojan.boqx	13	18	0.50	1.00	1.00	0.75
android.trojan.killav	14	17	1.00	1.00	1.00	1.00
android.riskware.tocrenu	15	14	0.50	1.00	1.00	0.75
android.exploit.gingerbreak	16	12	1.00	1.00	1.00	1.00
android.trojan.bankun	17	12	1.00	1.00	1.00	1.00
android.trojan.smsspy	18	11	1.00	1.00	1.00	1.00

3.1   Online Classification Algorithms

In general, an online learning algorithm works in a sequence of consecutive rounds. At round t, the algorithm takes an instance $x_t \in \mathbb{R}^d$, d-dimensional vector, as input to make the prediction $\hat{y}_t \in \{+1, -1\}$ (for binary classification) regarding to its current prediction model. After predicting, it receives the true label $y_t \in \{+1, -1\}$ and updates its model (a.k.a. hypothesis) based on prediction loss $\ell(y_t, \hat{y}_t)$ meaning the incompatibility between prediction and actual class. The goal of online learning is to minimize the total number of incorrect predictions; $sum(t : y_t \neq \hat{y}_t)$. Pseudo-code for generic online learning is given in Algorithm-1.

3.2   Classification Metrics

To evaluate the proposed method, the following class-specific metrics are used: *precision*, *recall* (a.k.a. sensitivity), *specificity*, *balanced accuracy*, and *overall accuracy* (the overall correctness of the model). Recall is the probability for a sample in class c to be classified correctly. On the contrary, specificity is

---

Algorithm 1. Generic online learning algorithm

Input : $\boldsymbol{w}_{t=1} = (0, ..., 0)$
1 **foreach** *round t in (1,2,..,N)* **do**
2 | Receive instance $\boldsymbol{x}_t \in \mathbb{R}^d$
3 | Predict label of $\boldsymbol{x}_t$: $\hat{y}_t = sign(\boldsymbol{x}_t.\boldsymbol{w}_t)$
4 | Obtain true label of the $\boldsymbol{x}_t$: $y_t \in \{+1, -1\}$
5 | Calculate the loss: ℓ_t
6 | Update the weights: $\boldsymbol{w}_{t+1}$
7 **end**
    Output: $\boldsymbol{w}_{t=N} = (w_1, ..., w_d)$

---

the probability for a sample not in class c to be classified correctly. The metrics are given as follows:

$$precision = \frac{tp}{tp + fp} \tag{1}$$

$$recall = \frac{tp}{tp + fn} \tag{2}$$

$$specificity = \frac{tn}{tn + fp} \tag{3}$$

$$balanced\ accuracy = \frac{recall + specificity}{2} = \frac{1}{2}\left(\frac{tp}{tp + fn} + \frac{tn}{tn + fp}\right) \tag{4}$$

$$accuracy = \frac{correctly\ classified\ instances}{total\ number\ of\ instances} \tag{5}$$

For instance, consider a given class c. True positives (tp) refer to the number of the samples in class c that are correctly classified while true negatives (tn) are the number of the samples not in class c that are correctly classified. False positives (fp) refer the number of the samples not in class c that are incorrectly classified. Similarly, false negatives (fn) are the number of the samples in class c that are incorrectly classified. The terms positive and negative indicate the classifier's success, and true and false denotes whether or not the prediction matches with ground truth label.

3.3 Testing Accuracy Results

The accuracy of testing is computed subject to different value of regularization weight parameter. The regularization weight parameter is denoted by C and determines the size of weight change at each iteration. A larger value means a possibility of a higher change in the updated weight vector and the model is created faster. But as a consequence, the model becomes more dependent to the training set and more susceptible to noise data. 10-fold cross-validation approach is used. The class-wise results for the most successful algorithm (i.e. Confidence-weighted linear classification in [15]) according to the different weight C are given in Table 5.

Table 5. Classification accuracy versus different regularization weight parameter

$C = 1$	$C = 2$	$C = 3$	$C = 4$	$C = 5$	$C = 10$	$C = 100$
0.81	0.83	0.84	0.89	0.80	0.78	0.76

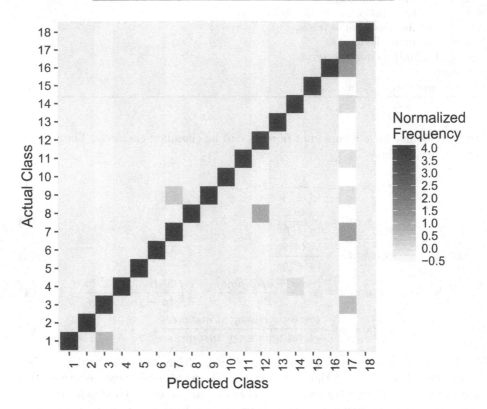

Fig. 1. Normalized confusion matrix

To analyze how well the classifier can recognize instance of different classes, we created the confusion matrix as shown in Fig. 1. The confusion matrix displays the number of correct and incorrect predictions made by the classifier with respect to ground truth (actual classes). The diagonal elements in the matrix represent the number of correctly classified instances for each class, while the off-diagonal elements represent the number of misclassified elements by the classifier. The higher the diagonal values of the confusion matrix are, the better the model fits the dataset (higher accuracy in individual family prediction). Since android.trojan.bankun family combines many functionalities executed also by other families in our dataset, android.trojan.agent, android.trojan.smskey and android.exploit.gingerbreak are incorrectly estimated as android.trojan.bankun.

4   Conclusions

This paper addresses the challenge of classifying android malware samples by using runtime artifacts while being robust to obfuscation. The presented classification system is usable on a large scale in real world due to its online machine learning methodology. The proposed method uses run-time behaviors of an executable to build the feature vector. We evaluated an online machine learning algorithm with 2000 samples belonging to 18 families. The results of this study indicate that runtime behavior modeling is a useful approach for classifying an android malware.

Acknowledgments. The authors gratefully acknowledge the support of Galatasaray University, scientific research support program under grant #16.401.004.

References

1. Internet Security Threat Report (2016) Available via Symantec. https://www.symantec.com/content/dam/symantec/docs/reports/istr-21-2016-en.pdf. Cited 15 Jun 2016
2. Schmidt, A.D., Bye, R., Schmidt, H.G., Clausen, J., Kiraz, O.: Static analysis of executables for collaborative malware detection on Android. In: 2009 IEEE International Conference on Communications, Dresden, pp. 1–5 (2009)
3. Peiravian, N., Zhu, X.: Machine learning for android malware detection using permission and API calls. In: Proceedings of the ICTAI 2013, The IEEE 25th International Conference on Tools with Artificial Intelligence, pp. 300–305 (2013)
4. Aung, Z., Zaw, W.: Permission-based android malware detection. Int. J. Scient. Technol. Res. **2**, 228–234 (2013)
5. Dhaya, R., Poongodi, M.: Detecting software vulnerabilities in android using static analysis. In: Proceedings of ICACCCT, Communication IEEE International Conference on Advanced Communication Control and Computing Technologies, pp. 915–918 (2014)
6. Tangil, G.S., Tapiador, J.E., Lopez, P.P., Blasco, J.: A text mining approach to analyzing and classifying code structures in android malware families. Expert Syst. Appl. **4**, 1104–1117 (2014)
7. Apvrille, A., Strazzere, T.: Reducing the window of opportunity for Android malware gotta catch em all. J. Comput. Virol. **8**, 61–71 (2012)

8. Xu, K., Li, Y., Deng, R.H.: ICCDetector: ICC-based malware detection on Android. Inf. Forensics Sec. **11**, 1252–1264 (2016)
9. Abdelrahman, O.H., Gelenbe, E., Görbil, G., Oklander, B.: Mobile network anomaly detection and mitigation: the NEMESYS approach. In: Gelenbe, E., Lent, R. (eds.) Information Sciences and Systems. LNEE, vol. 264, pp. 429–438. Springer, Switzerland (2013). doi:10.1007/978-3-319-01604-7_42
10. Gelenbe, E., Görbil, G., Tzovaras, D., Liebergeld, S., Garcia, D., Baltatu, M., Lyberopoulos, G.: NEMESYS: enhanced network security for seamless service provisioning in the smart mobile ecosystem. In: Information Sciences and Systems (2013). doi:10.1007/978-3-319-01604-7_36
11. Gelenbe, E., Görbil, G., Tzovaras, D., Liebergeld, S., Garcia, D., Baltatu, M., Lyberopoulos, G.: Security for smart mobile networks: the NEMESYS approach. In: Proceedings of the Global High Tech Congress on Electronics, pp. 63–69. IEEE (2013)
12. Cuckoo Sandbox (2016). cuckoosandbox.org. Cited 15 Jun 2016
13. Virusshare: Malware Sharing Platform (2016). http://www.virusshare.com/
14. Virustotal: An online multiple AV Scan Service (2016). http://www.virustotal.com/
15. Dredze, M., Crammer, K., Pereira, F.: Confidence-weighted linear classification. In: Proceedings of the 25th International Conference on Machine Learning, pp. 264–271. ACM (2008)

Comparison of Cross-Validation and Test Sets Approaches to Evaluation of Classifiers in Authorship Attribution Domain

Grzegorz Baron[✉]

Silesian University of Technology, Akademicka 16, 44-100 Gliwice, Poland
grzegorz.baron@polsl.pl

Abstract. The presented paper addresses problem of evaluation of decision systems in authorship attribution domain. Two typical approaches are cross-validation and evaluation based on specially created test datasets. Sometimes preparation of test sets can be troublesome. Another problem appears when discretization of input sets is taken into account. It is not obvious how to discretize test datasets. Therefore model evaluation method not requiring test sets would be useful. Cross-validation is the well-known and broadly accepted method, so the question arose if it can deliver reliable information about quality of prepared decision system. The set of classifiers was selected and different discretization algorithms were applied to obtain method invariant outcomes. The comparative results of experiments performed using cross-validation and test sets approaches to system evaluation, and conclusions are presented.

1 Introduction

Evaluation of classifier or classifiers applied in a decision system is the important step during a model building process. Two approaches are typical: cross-validation and using of test datasets. Both have some advantages and disadvantages. Cross-validation is easy to apply and in different application domains is accepted as good tool for measuring of classifiers performance. Evaluation based on test datasets requires at the beginning preparation of special sets containing data disjunctive of training one used during the creation process of a decision system. Sometimes it can be difficult to satisfy such condition.

Another issue, which arose during the author's former research, was utilization of test sets in conjunction with discretization of input data [3]. There are fundamental questions, how discretize test datasets in relation to learning sets to keep both sets coherent. Some approaches were analyzed, but they did not deliver unequivocal results. Therefore another idea came out - use of cross-validation instead of test data to validate the decision system. Such approach required deeper investigation and comparison with the first method of model validation. The paper presents experimental results, discussion and conclusions about that issue.

Authorship attribution is a part of stylometry which deals with recognition of texts' authors. Subject of analysis ranges from short Twitter messages to huge

© The Author(s) 2016
T. Czachórski et al. (Eds.): ISCIS 2016, CCIS 659, pp. 81–89, 2016.
DOI: 10.1007/978-3-319-47217-1_9

works of classical writers. Machine learning techniques and statistic-oriented methods are mainly involved in that domain. Different authorship attribution tasks have been categorized in [12], and three kinds of problems were formulated: profiling – there is no candidate proposed as an author; the needle-in-a-haystack – author of analyzed text should be selected from thousands of candidates; verification – there is an candidate to be verified as author of text.

The first important issue is to select characteristic features (attributes) to obtain author invariant input data which ensure good quality and performance of decision system [16]. Linguistic or statistical methods can be applied for that purpose. The analysis of syntactic, orthographic, vocabulary, structure, and layout text properties can be performed in that process [9].

The next step during building a decision system for authorship attribution task is selecting and applying the classifier or classifiers. Between different methods some unsupervised ones like cluster analysis, multidimensional scaling and principal component analysis can be mentioned. Supervised algorithms are represented by neural networks, decision trees, bayesian methods, linear discriminant analysis, support vector machines, etc. [9,17]

As aforementioned the aim of presented research was to compare two general approaches to evaluation of decision system: cross-validation [10] and test datasets utilization. To obtain representative results, a set of classifiers was chosen, applied and tested for stylometric data performing authorship attribution tasks. The idea was to select classifiers characterized by different ways of data processing. Finally the following suite of classifiers was applied: Naive Bayes, decision tree C4.5, k-Nearest Neighbors k-NN, neural networks – multilayer perceptron and Radial Basis Function network – RBF, PART, Random Forest. Test were performed for non-discretized and discretized data applying different approaches to test datasets discretization [3].

The paper is organized as follows. Section 2 presents the theoretical background and methods employed in the research. Section 3 introduces the experimental setup, datasets used and techniques employed. The test results and their discussion are given in Sect. 4, whereas Sect. 5 contains conclusions.

2   Theoretical Background

The main aims of presented research were analysis and comparison of cross-validation and test dataset approaches to evaluation of classifier or classifiers used in decision system especially in authorship attribution domain. Therefore a suite of classifiers has been set. The main idea was to select classifiers which behave differently because of performed algorithm and way of data processing. The final list of used classifiers contains: decision trees – PART [6] and C4.5 [14], Random Forest [4], k-Nearest Neighbors [1], Multilayer Perceptron, Radial Basis Function network, Naive Bayes [8].

Discretization is a process which allows to change the nature of data – it converts continuous values into nominal (discrete) ones. Two main circumstances can be mentioned, where discretization may or even must be applied. The first situation is when there are some suspicions about possible improvement of a decision system quality when discretized data is applied [2]. The second one is when method or algorithm employed in decision system can operate only on nominal, discrete data.

Because discretization reduces amount of data to be processed in a subsequent modules of decision system, sometimes it allows to filter information noise or allow to represent data in more consistent way. But on the other hand improper discretization application can lead to significant loss of information, and to degradation of overall performance of decision system.

Discretization algorithms can be divided basing on the different criterions. There are global methods which operate on whole attribute domain or local ones which process only part of input data. There are supervised algorithms which utilize class information in order to select bin ranges more accurately or unsupervised ones which perform only basic splitting of data into desired number of intervals [13]. Unsupervised methods are easier in implementation but supervised ones are considered to be better and more accurate.

In the presented research four discretization methods were used: equal width binning, equal frequency binning, as representatives of unsupervised algorithms, and supervised Fayyad & Irani's MDL [5] and Kononenko MDL [11].

The equal width algorithm divides the continuous range of a given attribute values into required number of discrete intervals and assigns to each value a descriptor of appropriate bin. The equal frequency algorithm splits the range of data into a required number of intervals so that every interval contains the same number of values.

During the developing of decision system, where input data is discretized and classifier is evaluated using test datasets, another question arises, namely how to discretize test datasets in relation to training data. Depending on the discretization methods different problems can appear such as uneven number of bins in training and test data, or cut-points which define boundaries of bins can be different in both datasets. That can lead to some inaccuracy during the evaluation of decision system. In [3] three approaches to discretization of test datasets were proposed:

- "independent" (Id) training and test datasets are discretized separately,
- "glued" (Gd) – training and test datasets are concatenated, the obtained set is discretized, and finally resulting dataset is split back into learning and test sets,
- "test on learn" (TLd) – firstly training dataset is discretized, and then test set is processed using cut-points calculated for training data.

3 Experimental Setup

The following steps were performed during the execution of experiments:

1. training and test data preparation,
2. discretization of input data applying selected algorithms using various approaches to test data processing,

3. training of selected classifiers,
4. system evaluation using cross-validation and test data approaches.

Input datasets were built basing on the several works of two male and two female authors. To obtain input data containing characteristic features satisfying author invariant requirement the following procedure was employed. Some linguistic descriptors from lexical and syntactic groups were chosen [15]. The works of each author were divided into parts. Then for each part frequencies of usages of selected attributes were calculated. Finally separate training and test sets were prepared with two classes (corresponding to two authors) in each. Attention was given during data preparation in order to obtain well-balanced training sets.

All experiments were performed using WEKA workbench, especially discretization methods and classifiers come from that software suite. It was necessary to make some modifications and develop additional methods to implement discretization algorithms allowing to discretize test data in "test on learn" and "glued" manner. Unsupervised discretization such as equal width and equal frequency were performed for required number of bins parameter ranged from 2 to 10. Base on the author's former experiences that was the range, where results are worth of notice.

According to the main aim of the presented research classifiers were evaluated using cross-validation and test datasets. Cross-validation was performed typically in 10-folds version. As a measure of classifier quality the number of correctly classified instances was taken.

4   Results and Discussion

The experiments were performed separately for male and female authors but final results were averaged for analysis and presentation purposes. For both neural network classifiers the best results obtained during experiments performed using multistart strategy are presented. Abbreviations used for classifiers naming in Figs. 1–3 are as follows: NB – Naive Bayes, C4.5 – decision tree C4.5, Knn – k-Nearest Neighbors, PART – decision tree PART, RF – Random Forest, RBF – Radial Basis Function network, MLP – Multilayer Perceptron. Additionally in Fig. 3 postfix "_T" denotes results obtained for evaluation using test data whereas postfix "_CV" is used for cross-validation results.

Results of the preliminary experiments performed for non-discretized data are presented in Fig. 1. It is easy to notice that classifiers performance measured using cross-validation are about 10 % better than results obtained for evaluation performed using test datasets. Only k-Nearest Neighbor classifier behave slightly better for evaluation using test data.

Figure 2 shows comparative results obtained for both analyzed evaluation approaches for data discretized using Kononenko MDL and Fayyad & Irani MDL respectively. Because test datasets were discretized using "Test on Learn", "Glued", and "Independent" approaches, the X axis is parted into three sections

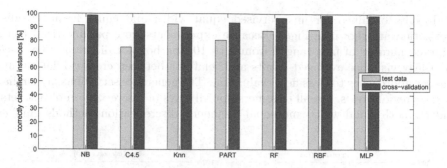

Fig. 1. Performance of classifiers for non-discretized data for evaluation performed using cross-validation and test datasets

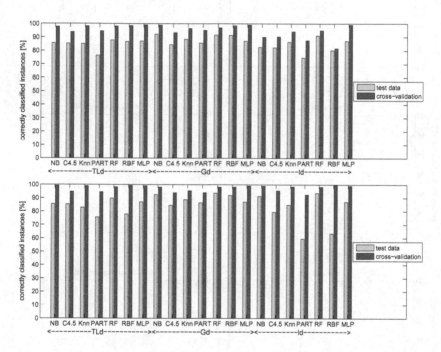

Fig. 2. Performance of classifiers for data discretized using supervised Kononenko MDL (above) and Fayyad & Irani MDL (below) for evaluation performed using cross-validation and test datasets. Three sections of the X axis present evaluation results obtained for test datasets discretized using "Test on Learn" – TLd, "Glued" – Gd, and "Independent" – Id approaches

which present results for mentioned ways of discretization. The huge domination of outcomes obtained for cross-validation evaluation is visible. Especially for "Independent" discretization of test datasets differences are big for PART and RBF classifiers.

Results obtained for unsupervised equal width and equal frequency discretization are shown in Fig. 3. Because experiments were parametrized using required number of bins ranged from 2 to 10, the boxplot diagrams were used to clearly visualize averaged results and relations between cross-validation and test set approaches to classifiers evaluation. The general observations are similar to the previous ones. For all classifiers, for all ways of discretization of test sets, and for both equal width and equal frequency discretization methods number

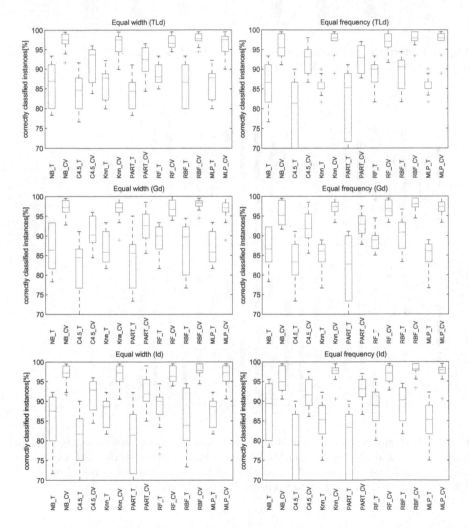

Fig. 3. Performance of classifiers for data discretized using unsupervised equal width (left column) and equal frequency (right column) discretization performed using the following approaches: "Test on Learn" – TLd (top row), "Glued" – Gd (middle row), and "Independent" – Id (bottom row), for evaluation performed using cross-validation ("_CV") and test datasets ("_T")

of correctly classified instances reported for cross-validation evaluation is bigger than for test dataset approach. The average difference is about 10 % (taking the medians of boxplots as reference points).

Summarizing the presented observations it can be stated that for almost all experiments (only one exception was observed) evaluation performed using cross-validation delivered quality measurements about 10 % greater comparing to the evaluation based on test datasets. In some cases that results reached 100 %. This is a problem because can lead to false conclusions about real quality of created decision system. Practically it is impossible to develop a system working with so high efficiency. Evaluation based on test datasets proved this opinion. Test sets were prepared basing on the texts other than that used for training of classifiers. So that evaluation results can be considered as more reliable. Depending on the classifier and discretization method they are smaller up to 30 %.

The general conclusion is that cross-validation which is acceptable and broadly used in different application domains is rather not useful for evaluating of decision systems in authorship attribution tasks performed in conditions and for data similar to that presented in the paper. If one decides to apply this method, must take into account that real performance of the system is much worse than reported using cross-validation evaluation.

5 Conclusions

The paper presents research on evaluation of decision systems in authorship attribution domain. Two typical approaches, namely cross-validation and evaluation based on specially created test datasets are considered. The research was the attempt to answer the question if evaluation using test datasets can be replaced by cross-validation to obtain reliable information about overall decision system quality. The set of different classifiers was selected and different discretization algorithms were applied to obtain method invariant outcomes. The comparative results of experiments performed using cross-validation and test sets approach to system evaluation are shown.

For almost all experiments (there were only one exception) evaluation performed using cross-validation delivered quality measurements (percent of correctly classified instances) about 10 % greater comparing to the evaluation based on test datasets. There were outliers where difference up to 30 % could be observed. On the other hand in some cases number od correctly classified instances for cross-validation was equal to 100 % what is not probable in real live tasks.

Concluding the research, it must be stated that cross-validation is rather not useful method for evaluating of decision systems in authorship attribution domain. It can be conditionally applied but strong tendency to overrating the quality of examined decision system must be taken into consideration.

Acknowledgments. The research described was performed at the Silesian University of Technology, Gliwice, Poland, in the framework of the project BK/RAu2/2016. All experiments were performed using WEKA workbench [7] basing on texts downloaded from http://www.gutenberg.org/.

References

1. Aha, D.W., Kibler, D., Albert, M.K.: Instance-based learning algorithms. In: Machine Learning, pp. 37–66 (1991)
2. Baron, G.: Influence of data discretization on efficiency of Bayesian Classifier for authorship attribution. Procedia Comput. Sci. **35**, 1112–1121 (2014)
3. Baron, G., Harezlak, K.: On Approaches to discretization of datasets used for evaluation of decision systems. In: Czarnowski, I., Caballero, A.M., Howlett, R.J., Jain, L.C. (eds.) Intelligent Decision Technologies 2016, vol. 57, pp. 149–159. Springer, Cham (2016)
4. Breiman, L., Schapire, E.: Random forests. In: Machine Learning, pp. 5–32 (2001)
5. Fayyad, U.M., Irani, K.B.: Multi-interval discretization of continuousvalued attributes for classification learning. In: 13th International Joint Conference on Articial Intelligence, vol. 2, pp. 1022–1027. Morgan Kaufmann Publishers (1993)
6. Frank, E., Witten, I.H.: Generating accurate rule sets without global optimization, pp. 144–151. Morgan Kaufmann (1998)
7. Hall, M., Frank, E., Holmes, G., Pfahringer, B., Reutemann, P., Witten, I.H.: The weka data mining software: an update. SIGKDD Explor. **11**(1), 10–18 (2009)
8. John, G., Langley, P.: Estimating continuous distributions in bayesian classifiers. In. Proceedings of the Eleventh Conference on Uncertainty in Artificial Intelligence, pp. 338–345. Morgan Kaufmann (1995)
9. Juola, P.: Authorship attribution. Found. Trends Inf. Retrieval **1**(3), 233–334 (2008)
10. Kohavi, R.: A study of cross-validation and bootstrap for accuracy estimation and model selection. In: International Joint Conference on Artificial Intelligence, pp. 1137–1143 (1995)
11. Kononenko, I.: On biases in estimating multi-valued attributes. In: 14th International Joint Conference on Articial Intelligence, pp. 1034–1040 (1995)
12. Koppel, M., Schler, J., Argamon, S.: Computational methods in authorship attribution. J. Am. Soc. Inform. Sci. Technol. **60**(1), 9–26 (2009)
13. Kotsiantis, S., Kanellopoulos, D.: Discretization techniques: a recent survey. Int. Trans. Comput. Sci. Eng. **1**(32), 47–58 (2006)

14. Quinlan, J.R.: C4.5: Programs for Machine Learning. Morgan Kaufmann Publishers Inc., San Francisco (1993)
15. Stańczyk, U.: Ranking of characteristic features in combined wrapper approaches to selection. Neural Comput. Appl. **26**(2), 329–344 (2015)
16. Stańczyk, U.: Establishing relevance of characteristic features for authorship attribution with ANN. In: Decker, H., Lhotská, L., Link, S., Basl, J., Tjoa, A.M. (eds.) DEXA 2013, Part II. LNCS, vol. 8056, pp. 1–8. Springer, Heidelberg (2013)
17. Stańczyk, U.: Rough set and artificial neural network approach to computational stylistics. In: Ramanna, S., Howlett, R.J. (eds.) Emerging Paradigms in ML and Applications. SIST, vol. 13, pp. 441–470. Springer, Heidelberg (2013)

Cosine Similarity-Based Pruning
for Concept Discovery

Abdullah Dogan[1]([⊠]), Alev Mutlu[2], and Pinar Karagoz[1]

[1] Department of Computer Engineering,
Middle East Technical University, Ankara, Turkey
{adogan,karagoz}@ceng.metu.edu.tr
[2] Department of Computer Engineering, Kocaeli University, Kocaeli, Turkey
alev.mutlu@kocaeli.edu.tr

Abstract. In this work we focus on improving the time efficiency of Inductive Logic Programming (ILP)-based concept discovery systems. Such systems have scalability issues mainly due to the evaluation of large search spaces. Evaluation of the search space cosists translating candidate concept descriptor into SQL queries, which involve a number of equijoins on several tables, and running them against the dataset. We aim to improve time efficiency of such systems by reducing the number of queries executed on a DBMS. To this aim, we utilize cosine similarity to measure the similarity of arguments that go through equijoins and prune those with 0 similarity. The proposed method is implemented as an extension to an existing ILP-based concept discovery system called Tabular Cris w-EF and experimental results show that the poposed method reduces the number of queries executed around 15 %.

1 Introduction

Concept discovery [3] is a multi-relational data mining task and is concerned with inducing logical definitions of a relation, called *target relation*, in terms of other provided relations, called *background knowledge*. It has extensively been studied under Inductive Logic Programming (ILP) [12] research and successful applications are reported [2,4,7,10].

ILP-based concept discovery systems consist of two main steps, namely *search space formation* and *search space evaluation*. In the first step candidate concept descriptors are generated and in the second step candiate condept descriptors are converted into queries, i.e. SQL queries, and are run against the dataset. As the search space is generally large and the queries involve multiple joins over several tables, the second step is computationally expensive and dominates the total running time of a concept discovery system. Several methods such as parallelization, memoization have been investigated to improve running time of the search space evaluation step.

In this paper we propose a method that improves the running time of concept discovery systems by reducing the number of SQL queries run on a database. The proposed method calculates the cosine similarity of the tables that appear

© The Author(s) 2016
T. Czachórski et al. (Eds.): ISCIS 2016, CCIS 659, pp. 90–96, 2016.
DOI: 10.1007/978-3-319-47217-1_10

in a query, and prunes those with 0 similarity. To realize this, (i) term-document count matrix where domain values of arguments of tables correspond to terms and relation arguments correspond to documents is built, and (ii) cosine similarity of table arguments that participate in a query are calculated from the term-document count matrix and those with 0 similarity are pruned.

The proposed method is implemented as an extension to an existing concept discovery system called Tabular CRIS w-EF [14,15]. To evaluate the performance of the proposed method several experiments are conducted on data sets that belong to different learning problems. The experimental results show that the proposed method reduces the number of queries executed by 15 % on the average without any loss in the accuracy of the systems.

The rest of the paper is organized as follows. In Sect. 2 we provide the background related to the study, in Sect. 3 we introduce the proposed method, and in Sect. 4 we present and discuss the experimental results. Last section concludes the paper.

2   Background

Concept discovery is a predictive multi relational data mining problem. Given a set facts, called *target instances*, and related observations, called *background knowledge*, concept discovery is concerned with inducing logical definitions of the target instances in terms of background knowledge. The problem has primarily been studied by ILP community and successful application have been reported.

In ILP-based concept discovery systems data is represented within first order logic framework and concept descriptors are generated by specialization or generalization of some an initial hypothesis. ILP-based concept discovery systems follow generate and test approach to find a solution and usually build large search spaces. Evaluation of the search space consists of translating concept descriptors into queries and running them against the data set. Evaluation of the queries is computationally expensive as queries involve multiple joins over tables. To improve running time of such systems several methods including parallelization [9], caching [13], query optimization [20] have been proposed. In parallelization based approaches either the search space is built or evaluated in parallel by multiple processors, in caching based methods queries and their results are stored in hash tables in case the same query is regenerated, and in query optimization based approaches several query optimization techniques are implemented to improve the running time of the search space evaluation step.

Cosine similarity is a popular metric to measure the similarity of data that can be represented as vectors. Cosine similarity of two vectors is the inner product of these vectors divided by the product of their lengths. Cosine similarity of -1 indicates exactly opposition, 1 indicates exact correlation, and 0 indicates decorrelation between the vectors. It has been applied in several domains including text document clustering [5], face verification [16].

In this work we propose to measure the cosine similarity of table arguments that partake in equijoins and prune those with cosine similarity of 0 without running them against the data set. To achieve this, firstly we group attributes that belong to the same domain, build a term-document matrix for each domain where domain values of the attributes constitute the terms, and individual arguments constitute the documents. When two arguments go through an equijoin we calculate their cosine similarity from the term-document matrix and prune those queries that have cosine similarity of 0. The proposed method is implemented as an extension to an existing ILP-based concept discovery system called Tabular CRIS w-EF. Tabular CRIS w-EF is an ILP-based concept discovery system that employs association rule mining techniques to find frequent and strong concept descriptors and utilizes memoization techniques to improve search space evaluation step of its predecessor CRIS [6].

3   Proposed Method

ILP-based systems represent the concept descriptors as Horn clauses where the positive literal represents the target relation, and the negated literals represent relations from the background knowledge. To evaluate such clauses, they are translated into SQL queries, where relations constitute the FROM clause and argument values form the WHERE clause of the query. As an example, consider the concept descriptor like *brother(A, B):-mother(C, A), mother(C, B)*. This concept descirptor is mapped to the following SQL query:

    SELECT SELECT b.arg1, brother.arg2
    FROM brother AS b, mother AS m1, mother AS m2
    WHERE brother.arg1=m1.arg2 and b.arg2=m2.arg2 and m1.arg1=m2.arg1

Fig. 1. Sample concept descriptor evaluation query

In such a transformation argument values with the same value go through equijoins. The proposed method targets such equijoins and prevents execution of queries that involve equjoins whose participating arguments have cosine similarity 0.

To achieve this,

(1) arguments are grouped based on their domains,
(2) for each such group term-document matrix is formed where values of the domain are the terms, arguments are the documents and values of an argument is the bag of the words of the argument
(3) for each term-document matrix a cosine similarity matrix is calculated.

To populate the count vector of an argument of a relation, i.e. *rel(arg1, ..., argn)* the following SQL statement is executed

ILP-based concept discovery systems construct concept descriptors in an iterative manner. At each iteration, a concept descriptor is specialized by appending

SELECT *arg*1, COUNT(*)-1 vector FROM
(SELECT *arg*1 FROM **rel**
UNION ALL
SELECT *arg*1 FROM **rel_domain**) t
GROUP BY *arg*1;

Fig. 2. Query for creating a count vector for rel.arg1

a new literal to the body of the concept descriptor in order to reduce the number of negative target instances it models, and it is evaluated. The proposed method inputs the refined concept descriptors, and checks if the newly added literal causes an equijoin. If and equijoin is detected, the cosine similarity of the arguments is fetched from the previously built matrix. If the cosine similarity is 0 then the concept descriptor is pruned, otherwise it is evaluated against the data set. If the newly added literal does not produce an equijoin then the query is directly evaluated against the data set. The proposed method is outlined in Algorithm 1.

Algorithm 1. PruneBasedOnSimilarity(vector<conceptDescriptors> C)

1: **for** (i = 0; i < C.size() ; i++) **do**
2:     newLiteral=C[C[i].literals.size()]
3:     **for** (j = 0; j < C[i].literals.size() - 1; j++) **do**
4:         **for** (k = 0; k < C[i].literals[j].arguments.size(); k++) **do**
5:             **for** (m = 0; m < newLiteral.argument.size(); m++) **do**
6:                 **if**   (C[i].literals[j].argument[k]=newLiteral.argument[m]   AND   similarity(C[i].literals[j].argument[k],newLiteral.argument[m])==0) **then**
7:                     prune pC[i]
8:                 **end if**
9:             **end for**
10:         **end for**
11:     **end for**
12: **end for**

In literature, there exists several ILP-based concept discovery systems that work on Prolog engines [11,17]. Such systems benefit from depth bounded interpreters for theorem proving to test possible concept descriptors. The proposed method is also applicable for such systems, as in Prolog notation each predicate can be considered a table and arguments of the literal can be considered as the fields of the table. With such a transformation, the proposed method can be utilized to prune hypotheses for ILP-based concept discovery systems that work on Prolog like environments.

In terms of algorithmic complexity, the proposed method consists of two main steps (i) matrix construction and (ii) cosine similarity calculation. To construct the matrix, one SQL query needs to be run for each literal argument. Complexity of cosine similarity is quadratic, hence applicable to real world data sets.

4   Experimental Results

To evaluate the performance of the proposed method we conducted experiments on data sets with different characteristics. Table 1 lists the data sets used in the experiments. Dunur and Elti are family relationship datasets. They are Turkish terms and are defined as follows: A is dunur of B if a child of A is married to a child of B, A is elti of B if As husband is brother of Bs husband. All the arguments of the two data sets belong to the same domain and both data sets are highly relational. Mutagenesis [19] and PTE [18] are biochemical datasets and aim is to classify the chemicals as to being related to mutagenicity and carcinogenicity or not, respectively. Mesh [1] is an engineering problem dataset where the problem is to find rules that define mesh resolution values of edges of physical structures. In the Eastbound [8] dataset there are two types of trains: (a) those that travel east called eastbound; and those that travel west called westbound. The problem is to find concept descriptors that define properties of the trains that travel to east. In these data sets there several domains that arguments belong to. The experiments are conducted on MySQL version 5.5.44-0ubuntu0.14.04.1. The DBMS resides on a machine with Core i7-2600K CPU processor and 7.8 GB RAM.

Table 1. Experimental parameters for each used data sets

Data set	Num of relations	Num of instances	Argument types
Dunur	9	234	Categorical
Elti	9	234	Categorical
Eastbound	12	196	Categorical, real
Mesh	26	1749	Categorical, real
Mutagenesis	8	16,544	Categorical, real
PTE	32	29,267	Categorical, real

In Table 2 we report the experimental results. Filtering Queries column shows the decrease in the number of queries when the proposed method is employed. The experimental results show that the proposed method performs well on the data sets that are highly relational, i.e. Dunur and Elti data sets. The proposed method performs sligly worse for the data sets that contains numerical attributes as well as categorical attributes to theose that only contains categorical attributes. This is indeed due to the fact that, arguments from the categorical domain go through equijoins, while arguments that belong to numerical domain go through less than ($<$), greater than ($>$) comparisons in SQL statements.

The last column of Table 2 reports the time impreovement when the proposed method is employed. When compared to decrease in the number of queries executed, the decrease in running time is less. This is due to the fact that Tabular CRIS w-EF employs advanced memoization mechanisms to store evaluation

Table 2. Improvements of proposed method

Data set	Tabular CRIS-wEF			Pruning by the proposed method			Improvement %		
	Num. rules	Num. queries	Time (mm:ss.sss)	Num. rules	Num. queries	Time (mm:ss.sss)	Rules	Queries	Time
Dunur	1887	5807	00:02.086	1279	4607	00:01.783	32.22	20.66	14.54
Elti	1741	5333	00:02.655	1422	4922	00:02.470	18.32	7.71	6.99
Eastbound	7294	34654	00:04.091	6805	32665	00:03.895	6.70	5.74	4.77
Mesh	56512	249084	00:27.302	54314	238982	00:27.314	3.89	4.06	−0.05
Mutagenesis	62486	223644	34:04.099	55477	216635	33:42.752	11.22	3.13	1.04
PTE	64322	237082	35:50.340	58503	231191	35:15.975	9.05	2.48	1.60
PTE No Aggr.	11166	43862	03:46.457	10328	43024	03:40.578	7.50	1.91	2.60

queries and retrieve results of repeated queries from hash tables. Nevertheless, the proposed method improves the running time of Tabular CRIS w-EF around 7.5 % on average.

5 Conclusion

Concept discovery systems face scalability issues due to the evaluation of the large search spaces they build. In this paper we propose a pruning mechanism based on cosine similarity to improve running time of concept discovery systems. The proposed method calculates the cosine similarity of arguments that participate in equijoins and prunes those concept descriptors that have arguments with cosine similarity 0. The proposed method is applicable to concept descovery systems that work on relational databases or Prolog like engines. The experimental results show that the proposed method decreased the number of concept descriptor evaluations around 15 % on the average, and improved the running time of the system around 7.5 % on the average.

References

1. Dolšak, B.: Finite element mesh design expert system. Knowl. Based Syst. **15**(5), 315–322 (2002)

2. Dolsak, B., Muggleton, S.: The application of inductive logic programming to finite element mesh design. In: Inductive Logic Programming, pp. 453–472. Academic Press (1992)
3. Dzeroski, S.: Multi-relational data mining: an introduction. SIGKDD Explor. **5**(1), 1–16 (2003). doi:10.1145/959242.959245
4. Feng, C.: Inducing temporal fault diagnostic rules from a qualitative model. In: Proceedings of the Eighth International Workshop (ML91), Northwestern University, Evanston, Illinois, USA, pp. 403–406 (1991)
5. Huang, A.: Similarity measures for text document clustering. In: Proceedings of the Sixth New Zealand Computer Science Research Student Conference (NZC-SRSC2008), Christchurch, New Zealand, pp. 49–56 (2008)
6. Kavurucu, Y., Senkul, P., Toroslu, I.H.: ILP-based concept discovery in multi-relational data mining. Expert Syst. Appl. **36**(9), 11418–11428 (2009). doi:10.1016/j.eswa.2009.02.100
7. King, R.D., Muggleton, S., Lewis, R.A., Sternberg, M.: Drug design by machine learning: the use of inductive logic programming to model the structure-activity relationships of trimethoprim analogues binding to dihydrofolate reductase. Proc. Nat. Acad. Sci. **89**(23), 11322–11326 (1992)
8. Larson, J., Michalski, R.S.: Inductive inference of VL decision rules. ACM SIGART Bull. **63**, 38–44 (1977)
9. Matsui, T., Inuzuka, N., Seki, H., Itoh, H.: Comparison of three parallel implementations of an induction algorithm. In: 8th International Parallel Computing Workshop, pp. 181–188. Citeseer (1998)
10. Muggleton, S., King, R., Sternberg, M.: Predicting protein secondary structure using inductive logic programming. Protein Eng. **5**(7), 647–657 (1992)
11. Muggleton, S.: Inverse entailment and progol. New Gener. Comput. **13**(3–4), 245–286 (1995)
12. Muggleton, S., Raedt, L.D.: Inductive logic programming: theory and methods. J. Log. Program. **19**(20), 629–679 (1994). doi:10.1016/0743-1066(94)90035-3
13. Mutlu, A., Karagoz, P.: Policy-based memoization for ILP-based concept discovery systems. J. Intell. Inf. Syst. **46**(1), 99–120 (2016). doi:10.1007/s10844-015-0354-7
14. Mutlu, A., Senkul, P.: Improving hash table hit ratio of an ILP-based concept discovery system with memoization capabilities. In: Gelenbe, E., Lent, R. (eds.) Computer and Information Sciences III, pp. 261–269. Springer, London (2012). doi:10.1007/978-1-4471-4594-3_27
15. Mutlu, A., Senkul, P.: Improving hit ratio of ILP-based concept discovery system with memoization. Comput. J. **57**(1), 138–153 (2014). doi:10.1093/comjnl/bxs163
16. Nguyen, H.V., Bai, L.: Cosine similarity metric learning for face verification. In: Kimmel, R., Klette, R., Sugimoto, A. (eds.) ACCV 2010. LNCS, vol. 6493, pp. 709–720. Springer, Heidelberg (2011). doi:10.1007/978-3-642-19309-5_55
17. Quinlan, J.R.: Learning logical definitions from relations. Mach. Learn. **5**(3), 239–266 (1990)
18. Srinivasan, A., King, R.D., Muggleton, S.H., Sternberg, M.J.: The predictive toxicology evaluation challenge. In: IJCAI, vol. 1, pp. 4–9. Citeseer (1997)
19. Srinivasan, A., Muggleton, S.H., Sternberg, M.J., King, R.D.: Theories for mutagenicity: a study in first-order and feature-based induction. Artif. Intell. **85**(1), 277–299 (1996)
20. Struyf, J., Blockeel, H.: Query optimization in inductive logic programming by reordering literals. In: Horváth, T., Yamamoto, A. (eds.) ILP 2003. LNCS (LNAI), vol. 2835, pp. 329–346. Springer, Heidelberg (2003). doi:10.1007/978-3-540-39917-9_22

A Critical Evaluation of Web Service Modeling Ontology and Web Service Modeling Language

Omid Sharifi[1] and Zeki Bayram[2(✉)]

[1] Computer and Software Engineering Department, Toros University, Mersin, Turkey
omid.sharifi@toros.edu.tr
[2] Computer Engineering Department, Eastern Mediterranean University,
Famagusta, Cyprus
zeki.bayram@emu.edu.tr

Abstract. Web Service Modeling Language (WSML), based on the Web
Service Modeling Ontology (WSMO), is a large and highly complex language designed for the specification of semantic web services. It has different variants based on logical formalisms, such as Description Log-
ics, First-Order Logic and Logic Programming. We perform an in-depth
study of both WSMO and WSML, critically evaluating them by identifying their strong points and areas in which improvement would be
beneficial. Our studies show that in spite of all the features WSMO and
WSML support, their sheer size and complexity are major weaknesses,
and there are other areas in which important deficiencies exist as well.
We point out those discovered deficiencies, and propose remedies for
them, laying the foundation for a more tractable and useful formalism
for specifying semantic web services.

Keywords: Semantic web services · WSMO · WSML · Evaluation

1 Introduction

The goal of web services is to allow normally incompatible applications to inter-
operate over the Web regardless of language, platform, or operating system [10].
Web services are much like remote procedure calls, but they are invoked using
Internet and WWW standards and protocols such as Simple Object Access Protocol (SOAP) [2] and Hypertext Transfer Protocol (HTTP) [1].

Web Services Modeling Ontology (WSMO) [3] is a comprehensive framework
for describing web services, goals (high-level queries for finding web services),
mediators (mappings for resolving heterogeneities) and ontologies. Web Services
Modeling Language (WSML) [5] is a *family* of concrete languages based on F-
logic [11] that implement the WSMO framework. The variants of WSML are
WSML-core, WSML-flight, WSML-rule, WSML-DL, and WSML-full. WSML is
large, relatively complex, and somewhat confusing, with different variants being
based on different formalisms. The complexity and confusion arise mainly from
the many variants of the language, and the rules used to define the variants.

© The Author(s) 2016
T. Czachórski et al. (Eds.): ISCIS 2016, CCIS 659, pp. 97–105, 2016.
DOI: 10.1007/978-3-319-47217-1_11

The variants of WSML form a hierarchy, with WSML-full being on top (the most powerful) and WSML-core being at the bottom (weakest).

Our literature search has failed to reveal any significant industrial real-life application that uses WSML. We believe this is due to the inherent complexity of the language, the "less-than-complete" state of WSML (e.g. the syntax of WSML-DL does not conform to the usual description logic syntax, choreography specification using abstract state machines (ASM) [8] seems unfit for the job due to the execution semantics of ASMs, goals, choreographies and web services are not integrated in the same logical framework etc.), as well as the lack of proper development tools and execution environments. So WSML looks like it is still in a "work-in-progress" state, rather than a finished product.

In this work, we critically evaluate the strengths and weaknesses of WSMO and WSML, and determine the areas of improvement that will result in a usable semantic web service specification language. This is the main contribution of this work, which will be input to the next phase of our research, the actual design and implementation of such a language.

The remainder of the paper is organized as follows. Section 2 contains a critical evaluation of WSMO and WSML, including their strengths, weaknesses and deficiencies, discovered both through our detailed study of the documentation provided for WSMO and WSML, as well as experimentation with the paradigm in several use-cases. In Sect. 3 we have a brief discussion of related work, and finally Sect. 4 is the conclusion and future research directions.

2   Evaluation of WSMO and WSML

In this section we discuss the strong and weak points of WSMO and WSML as discovered through our studies of their specification and the practical experience gained through experimentation. We also suggest possible improvements wherever possible.

2.1   General Observations

WSMO boasts a comprehensive approach that tries to leave no aspect of semantic web services out. These include ontologies, goals, web services and mediators. In the same spirit of thoroughness, designers of WSML have adopted the paradigm of trying to provide everything everybody could ever want and let each potential user chose the "most suitable" variant of the language for the job at hand. This approach has resulted in a complex syntax, as well as a complex set of rules that differentiate one version of the language form another.

2.2   Deficiencies in Syntax

WSML-DL and WSML-full have no explicit syntax for the description logic component [5], relying on a first-order encoding of description logic statements. Without proper syntax, it is not possible to use them in the specification of semantic web services in a convenient way.

2.3 Logical Basis of WSMO

The ontology component of WSMO is based on F-logic, which gives this component a solid theoretical foundation. However, its precise relationship to F-logic has not been given formally, and what features of F-logic have been left out are not specified explicitly.

2.4 Lack of a Semantics Specification for Web Service Methods/Operations

In spite of all the effort at comprehensiveness, there are significant omissions in WSMO, such as specification of the semantics of actual methods (operations) that the web service provides, which makes it impossible to *prove* that after a "match" occurs between a goal and a web service, the post-condition of the goal will indeed be satisfied. Even worse, once matching succeeds and the web service is called according to the specified choreography, the *actual results* of the invocation may not satisfy the post-condition of the goal. Below, we explain why.

In WSMO, matching between a goal and web service occurs by considering the pre-post conditions of the goal and web service, and this is fine. The problem occurs because of the *lack* of a semantic specification (for example, in the form of pre-post conditions) for web service *methods/operations*, and how these methods are actually called through the execution of the choreography engine. Method calls are generated according to availability of "data" in the form of instances, and the mapping of instances to parameters of methods. There is no consideration of logical conditions which must be true before the method is called, and no guarantee of the state of the system after the method is called, since these are not specified for the web methods. Instances of a concept can be parameters to more than one web method. Assuming two methods A and B have the same signature, it may be the case that an unintended method call can be made to B, when in fact the call should have been made to A, which results in wrong computation. Consequently, not only is it impossible to *prove* that after a "match" occurs between a goal and a web service, the post-condition of the goal will be satisfied, but also once the web service execution is initiated, the computation itself can produce wrong results, invalidating the logical specification of the web service.

Unfortunately, the interplay between choreography, grounding and logical specification of what the web service does (including the lack of the specification of semantics for web service methods) has been overlooked in WSMO. All these components need development and integration in order to make them part of a *coherent* whole.

2.5 Implementation and Tool Support

Some developmental tools, such as the "Web Services Modeling Toolkit" [4] exist which make writing WSML specifications relatively easy. However, these tools depend on external reasoner support, rather than having intrinsic reasoning capabilities. As such, development and testing of semantic web service specifications cannot be made in a reliable manner. For example, no explanations are given when discovery fails for a given goal.

2.6 Choreography in WSMO

We have already talked about how the interplay of choreography and grounding can result in incorrect execution, invalidating the logical specification of a web service. In this section, we delve more deeply into the problems of WSMO choreography.

– WSMO choreography is purportedly based on the formalism of abstract state machines [8], but in fact it is only a crude approximation. Very significantly, evolving algebras are magically replaced with the state of the ontologies as defined by instances of relations and concepts. This transformation seems to have no logical basis, so the applicability of any theory developed for abstract state machines to WSMO choreography specifications is questionable. The choreography attempt of WSMO looks more like a forward-chained expert system shell, where the role of the "working memory" is played by the current set of instances in the ontologies. It probably would be more reasonable to consider WSMO choreography in this way, rather than being based on abstract state machines.
– The fact that in an abstract state machine rules are fired in parallel does not match well with the real life situation that method calls implied by the firing of rules have to be executed sequentially.
– Both goals and web services have choreography specifications, but there is no notion of how the choreographies of goals and web services are supposed to match during the discovery phase. It is also not clear how the two are supposed to interact during the execution phase. Although restrictions on who can modify the state of the ontology and in what way can be specified in the form of modes of concepts, this is relatively complex, and far from practical. In the documentation of WSMO, only the choreography of the service is made use of.
– Choreography grounding in WSMO tries to map instances to method parameters of the web service methods by relating concepts to the methods directly. Methods are then called when their parameters are available in the current working memory. The firing of the rules are intermixed with the invocation of methods (with appropriate lowering/lifting of parameters), and changes to working memory by actions on the right hand side are forbidden (presuming that any changes will be made by the actual method call). This is a strange state of affairs, since the client may itself need to add something to the working memory, and there is no provision for this.
– The choreography rule language allows nested rules. Although this nesting permits very expressive rules to be written, using the "if", "forall" and "choose" constructs in any combination in a nested manner, the resulting rules are prohibitively complex, both to understand, and to execute.

- As mentioned before, in the grounding process, only the availability of instances that can be passed as parameters to methods, and the pre-determined mapping between concepts and parameters, are considered, with no pre-conditions for method calls. This is a major flaw, since it may be that two methods have exactly the same parameter set, but they perform very different functions, and the wrong one gets called.
- The choreography specification is disparate from the capability specification (pre-conditions, post-conditions), whereas they are in fact intimately related and intertwined. The actions specified in the choreography should actually take the initial state of the ontologies to their final state, through the inter-action of the requester and web service. This fact is completely overlooked in WSMO choreography.
- Choreography engine execution stops in WSMO when no more rules apply. A natural time for it to stop would be when the conditions specified in the goal are satisfied by the current state of the ontology stores. Again this is a design flaw, which is due to the fact that the intimate relationship between the capability specification and choreography has been overlooked.

2.7 Orchestration in WSMO

The orchestration component of WSMO is yet to be defined. The creators of WSMO say it will be similar to choreography, and be part of the interface speci-fication of a web service. At a conceptual level, however, we find the specification of orchestration for a web service somewhat unnecessary. Why would a requester care about *how* a service provider provides its service? Composition of web ser-vices to achieve a goal *would* be much more meaningful, however. So the idea of placing orchestration within a web service specification seems misguided. Its proper place would be inside the specification of a *complex* goal, which would help and guide the service discovery component to not only find a service that meets the requirements of the goal, but also mix-and-match and compose differ-ent web services to achieve the requirements of the goal.

2.8 Goal Specification

The goal specification includes the components "assumptions," "pre-conditions," "post-conditions" and "effects," just like the web service specification. The logical correspondence between the "pre-conditions," "assumptions," "post-conditions" and "effects," of goals and web services is not specified at all. The usage of the same terminology for both goals and web services is also misleading. In reality, the web service *requires* that its pre-conditions and assumptions hold before it can be called, and *guarantees* that if it is called, the post-conditions and effects will be true. On the other hand, the goal declares that it *guarantees* a certain state, perhaps by adding instances to the instance store, of the world before it makes a request to a web service, and *requires* certain conditions to be true as a result of the execution of the web service. The syntax of the goals should be consistent with this state of affairs.

2.9   Reusing Goals Through Specialization

Being able to reuse an existing goal after specializing it in some way would be very beneficial. The template mechanism of programming languages, or "prepared queries with parameters" in the world of databases are concepts which can be adapted to goals in WSMO to achieve the required specialization. Such functionality is currently missing in WSML.

2.10   Specialization Mechanism for Web Service Specifications

Developing a web service specification from scratch is a very formidable task. Just like in the case of specializing goals, a mechanism for taking a "generic" web service specification in a domain, and specializing it to describe a specific web service functionality would be a very useful proposition. To take this idea even further, a hierarchy of web service specifications can be published in a central repository, and actual web services can just declare that they implement a pre-published specification in the hierarchy. Or, they can grow the hierarchy by specializing an existing specification, and "plugging" their specification into the existing hierarchy. Such an approach will help in service discovery as well. A specialization mechanism for web services does not exist in WSMO, and would be a welcome addition to it.

2.11   Missing Aggregate Function Capability

The logic used in WSML (even in WSML full) does not permit aggregate functions in the sense of database query languages (sum, average etc.). Such an addition however would require moving away from first order logic into higher order logic, with corresponding loss of computational tractability. Still, it may be worthwhile to investigate restricted classes of aggregate functionality which lend themselves to practical implementation. For example, a built-in *setof* predicate could be used to implement aggregate functions.

2.12   Extra-Logical Predicates

The ability to check whether a logic variable is bound to an object, or whether it is in an unbound state (the *var* predicate of Prolog [16]) is missing. The availability of this feature is of practical importance, since for example a web service pre-condition may be a disjunction, and depending on the input provided by the goal, some variables in the disjunction may remain unbound after a successful match.

2.13   Multiple Functionality in a Web Service

A WSML goal or web service may only have one capability [9]. This is a severe restriction, since a web service can possibly provide different results, depending on the provided input. Ideally, each web service specification should be able to have a *set* of capabilities. This is not currently available in WSMO or WSML.

2.14 Automatic Mapping Between Attributes and Relations

Although one can define a binary relation for each attribute using an axiom, relating objects and their attribute values, this is cumbersome when done manually. Having it done automatically would be nice, a feature currently not available in WSML.

2.15 Error Processing

There is currently no mechanism specifying how to handle errors when they arise. For example, what should be done when a constraint is violated in some ontology? There should be a way of communicating error conditions to the requester when they arise. This could be the counterpart of the exception mechanism in programming languages.

2.16 No Agreed-Upon Semantics for WSML-Full

WSML-full, which is a combination of WSML-DL and WSML-rule, has no agreed-upon semantics yet [9] yet. With no formal semantics available, it is hard to imagine how WSML-full specifications could be processed at all.

3 Related Work

The authors have benefited from practical experience gained through semantic web service specification use cases reported in [6,13,14] in determining weak points of WSMO and WSML, in addition to unreported extensive experimentation. Although some of the drawbacks of WSML reported here have been pointed out in the master thesis by Cobanoglu [7] as well, our coverage of the choreography issue is unique in its depth and scope. We also offer solutions wherever possible to improve WSMO and WSML.

Our literature search failed to reveal any additional comprehensive study on the weaknesses of WSMO and WSML. However, we should also mention WSMO-lite [12,15], a relatively recent bottom-up semantic web service specification framework inspired by WSMO, that recognizes and provides solutions for the problems of specifying pre and post conditions for web service operations, as well as dealing with error conditions.

4 Conclusion and Future Work

We investigated the WSMO semantic web service framework, and the WSML language through an in-depth study of both, as well as extensive practical experimentation. Our investigation has revealed several deficiencies and flaws with WSMO and WSML, which we presented in this paper. We also provided suggestions for improvement where possible.

In future work, we are planning to develop a logic based semantic web service framework that builds on the strengths of WSMO, but at the same time remedies the weaknesses identified in this paper. Our proposal will aim to be coherent, where all the components are in harmony with each other, manageable, not unnecessarily complex, and practical enough to be used in real life.

References

1. HTTP - hypertext transfer protocol. http://www.w3.org/Protocols/. Accessed 19 Apr 2016
2. SOAP version 1.2 part 1: Messaging framework (2nd edn.). https://www.w3.org/TR/soap12/. Accessed 19 Apr 2016
3. Web Service Modeling Ontology. http://www.wsmo.org/. Accessed 30 Mar 2016
4. Web Services Modelling Toolkit. https://sourceforge.net/projects/wsmt/. Accessed 18 Apr 2016
5. WSML - Web Service Modeling Language. http://www.wsmo.org/wsml. Accessed 30 Mar 2016
6. Çobanoğlu, Ş., Bayram, Z.: Semantic web services for university course registration. In: Kim, W., Ding, Y., Kim, H.-G. (eds.) JIST 2013. LNCS, vol. 8388, pp. 3–16. Springer, Heidelberg (2014)
7. Cobanoglu, S.: A critical evaluation of web service modeling language. Master Thesis, Eastern Mediterranean University, February 2013
8. Börger, E., Stärk, R.: Abstract State Machines: A Method for High-Level System Design and Analysis. Springer, Heidelberg (1984)
9. Group, W.W., et al.: D16.1v1.0 WSML language reference final draft 2008–08-08 (2008). http://www.wsmo.org/TR/d16/d16.1/v1.0/. Accessed 20 Apr 2016
10. McGovern, J., Tyagi, S., Stevens, M., Mathew, S.: Java Web Services Architecture. Morgan Kaufmann, San Francisco (2003)
11. Kifer, M., Lausen, G., Wu, J.: Logical foundations of object-oriented and frame-based languages. J. ACM **42**(4), 741–843 (1995). doi:10.1145/210332.210335
12. Roman, D., Kopeck, J., Vitvar, T., Domingue, J., Fensel, D.: WSMO-Lite and hRESTS: lightweight semantic annotations for web services and RESTful APIs. Web Semant. Sci., Serv. Agents WWW **31**, 39–58 (2015)
13. Sharifi, O., Bayram, Z.: Database modelling using WSML in the specification of a banking application. In: Proceedings WASET 2013, pp. 263–267 (2013)

14. Sharifi, O., Bayram, Z.: Specifying banking transactions using web services modeling language (WSML). In: Proceedings of the Fourth International Conference on Information and Communication Systems (ICICS 2013), pp. 138–143 (2013)
15. Vitvar, T., Kopecký, J., Viskova, J., Fensel, D.: WSMO-lite annotations for web services. In: Bechhofer, S., Hauswirth, M., Hoffmann, J., Koubarakis, M. (eds.) ESWC 2008. LNCS, vol. 5021, pp. 674–689. Springer, Heidelberg (2008). doi:10. 1007/978-3-540-68234-9_49
16. Clocksin, W.F., Mellish, C.S.: Programming in Prolog. Springer, Verlag (1984)

Weighting and Pruning of Decision Rules by Attributes and Attribute Rankings

Urszula Stańczyk[✉]

Silesian University of Technology, Akademicka 16, 44-100 Gliwice, Poland
urszula.stanczyk@polsl.pl

Abstract. Pruning is a popular post-processing mechanism used in search for optimal solutions when there is insufficient domain knowledge to either limit learning data or govern induction in order to infer only the most interesting or important decision rules. Filtering of generated rules can be driven by various parameters, for example explicit rule characteristics. The paper presents research on pruning rule sets by two approaches involving attribute rankings, the first relaying on selection of rules referring to the highest ranking attributes, which is compared to weighting of rules by calculated quality measures dependent on weights coming from attribute rankings that results in rule ranking.

Keywords: Decision rules · Pruning · Weighting · Attribute · Ranking

1 Introduction

Rule classifiers express patterns discovered in data in learning processes through conditions on attributes included in the premises and pointing to specific classes [5]. A variety of available approaches to induction enable construction of classifiers with minimal numbers of constituent rules, with all rules that can be inferred from the training samples, or with subsets of interesting elements [3].

To limit the number of considered rules [9] either pre-processing can be employed, with reducing rather data than rules, by selection of features or instances, or in-processing relaying on induction of only those rules that satisfy given requirements, or post-processing, which implements pruning mechanisms and rejection of some unsatisfactory rules. The paper focuses on this latter approach.

One of the most straightforward ways to prune rules and rule sets involves exploiting direct parameters of rules, such as their support, length [11], strength [1]. Also specific condition attributes can be taken into account and indicate rules to be selected by appearing in their premises [12]. Such process can lead to improved performance or structure and in the presented research it is compared to weighting of rules by calculated quality measures, also based on attributes [13], both procedures actively using rankings of considered characteristic features [7].

The paper is organised as follows. Section 2 briefly describes some elements of background, that is feature weighting and ranking, and aims of pruning of

© The Author(s) 2016
T. Czachórski et al. (Eds.): ISCIS 2016, CCIS 659, pp. 106–114, 2016.
DOI: 10.1007/978-3-319-47217-1_12

rules and rule sets. Section 3 explains the proposed research framework, details experimental setup, and gives test results. Section 4 concludes the paper.

2   Background

The research described in this paper incorporates characteristic feature weights and rankings into the problem of pruning of decision rules and rule sets.

2.1   Feature Ranking

Roles of specific features exploited in any classification task can vary in significance and relevance in a high degree. The importance of individual attributes can be discovered by some approach leading to their ranking, that is assigning values of a score function which causes putting them in a specific order [7].

Rankings of characteristic features can be obtained through application of statistical measures, machine learning approaches, or systematic procedures [12]. The former assign calculated weights to all variables, while the latter can return only the positions in a ranking, reflecting discovered order of relevance.

Information Gain coefficient (*InfoGain, IG*) is defined by employing the concept of entropy from information theory for attributes and classes:

$$InfoGain(Cl, a_f) = H(Cl) - H(Cl|a_f), \tag{1}$$

where $H(Cl)$ denotes the entropy for the decision attribute Cl and $H(Cl|a_f)$ condition entropy, that is class entropy while observing values of attribute a.

An attribute relevance measure can be based on rule length [11], with special attention given to the shortest rules that often possess good generalisation properties:

$$MREVM(a) = Nr(a, MinL) : Nr(a, MinL + 1), \tag{2}$$

where $Nr(a, L)$ denotes the number of rules with length L in which attribute a appears, and $MinL$ is the length of the shortest rule containing a. The attribute ranking constructed in this way is wrapped around the specific inducer, not its performance, since other parameters of rules are disregarded, but structure.

2.2   Pruning of Decision Rules

To limit the number of rules three approaches can be considered [8]:

- pre-processing — the input data is reduced before the learning stage starts by rejecting some examples or cutting down on characteristic features. With less data to infer from, it follows that fewer rules are induced.
- at the algorithm construction stage — by implementation of specific procedures only some rules meeting requirements are found instead of all possible.
- post-processing — the set of inferred rules is analysed and some of its elements discarded while others selected.

When lower numbers of rules are found the learning stage can be shorter, yet solutions are not necessarily the best. If higher numbers of rules are generated, more thorough and in-depth analysis is enabled, yet even for rule sets with small cardinalities some measures of quality or interestingness can be employed [6].

Rule quality can be weighted by conditional attributes [13]:

$$QM(r_i) = \prod_{j=1}^{K_{r_i}} w(a_j), \tag{3}$$

where K_{r_i} denotes the number of conditions included in rule r_i and $w(a_j)$ weight of a_j attribute taken from a ranking. It is assumed that $w(a_j) \in (0, 1]$.

3   Experimental Setup and Obtained Results

The research works presented were executed within the general framework:

- Initial preparation of learning and testing data sets
- Obtaining rankings of attributes
- Induction of decision algorithms
- Pruning of decision rules in two approaches:
  - Selecting rules referring to specific attributes in the ranking
  - Calculating measures for all rules while exploiting weights assigned to positions in the attribute rankings, which led to weighting of rules and their rankings, and from these rankings rules in turn were selected
- Comparison and analysis of obtained test results

Steps of these procedures are described in the following subsections.

3.1   Input Datasets

As a domain of application for the research stylometric analysis of texts was selected. Stylometry enables authorship attribution while basing on employed linguistic characteristic features. Typically they refer to lexical and syntactic markers, giving frequencies of occurrence for selected function words and punctuation marks that reflect individual habits of sentence and paragraph formation.

Learning and testing samples corresponded to parts of longer works by two pairs of writers, female and male, giving binary classification with balanced data.

As attribute values specified usage frequencies of textual descriptors, they were small fractions, which means that for data mining there was needed either some technique that can deal efficiently with continuous numbers, or some discretization strategy was required [2]. Since regardless of a selected method discretization always causes some loss of information, it was not attempted.

3.2   Rankings of Attributes

In the research presented two attribute rankings were tested. The first one relied on statistical properties detected in input datasets and was completely independent on the classifier used later for prediction, and the other was wrapped around characteristics of induced rules, observing how often each variable occurs in shortest rules, which usually are of higher quality as they are better at generalisation and description of detected patterns than those with many conditions. Orderings of variables for both rankings and both datasets are given in Table 1.

Table 1. Rankings of condition attributes

No	$w(a)$	Female writers		Male writers	
		InfoGain	*MREVM*	*InfoGain*	*MREVM*
1	1	not	not	and	and
2	1/2	:	:	that	by
3	1/3	;	but	by	from
4	1/4	,	and	but	of
5	1/5	-	.	from	in
6	1/6	on	,	what	:
7	1/7	?	by	for	!
8	1/8	(	for	-	on
9	1/9	as	to	?	,
10	1/10	but	this	if	as
11	1/11	by	as	at	(
12	1/12	that	what	with	with
13	1/13	for	!	not	
14	1/14	to	from	:	this
15	1/15	at	?	to	at
16	1/16	.	-	in	not
17	1/17	and	of	(	;
18	1/18	in	in	as	?
19	1/19	this	that	!	-
20	1/20	!	with	;	to
21	1/21	with	if	on	if
22	1/22	of	at	.	what
23	1/23	what	(	of	for
24	1/24	if	on	this	but
25	1/25	from	;	,	that

InfoGain returns a specific score for each feature while *MREVM* gives a ratio. To unify numbers considered as attribute weights they were assigned in an arbitrary manner, listed in column denoted $w(a)$, and equal $1/i$, where i is a position in the ranking. Thus the distances between weights decrease while going down the ranking. It is assumed that each variable has nonzero weight.

3.3 DRSA Rule Classifiers

The rules were induced with the help of 4eMka Software (developed at the Poznań University of Technology, Poland), which implements Dominance-Based Rough Set Approach (DRSA). By substituting the original indiscernibility relation [4] of classical rough sets with dominance DRSA observes ordinal properties in datasets and enables both nominal and ordinal classification [10].

As the reference points classification systems with all rules on examples were taken. For female writers the algorithm consisted of 62383 rules, which with constraints on minimal rule support to be equal at least 66 resulted in 17 decision rules giving the maximal classification accuracy of 86.67 %. For male writers the algorithm contained 46191 rules, limited to 80 by support equal at least 41, and it gave the correct recognition of 76.67 % of testing samples. In all cases ambiguous decisions were treated as incorrect, without any further processing.

3.4 Pruning of Rule Sets by Attributes

Selection of decision rules while following attribute rankings was executed as follows: at i-th step only the rules with conditions on the i highest ranking features were taken into account. The rules could refer to all or some proper subsets of variables considered, and these with at least one condition on any of lower ranking attributes were discarded. Thus at the first step only rules with single conditions on the highest ranking variable were filtered, while at the last 25-th step all features and all rules were included. For example at 5-th step for female writer dataset for *InfoGain* ranking only rules referring to any combination of attributes: not, colon, semicolon, comma, hyphen, were selected. The detailed results for both datasets and both rankings are listed in Table 2.

It can be observed that with each variable added to the studied set the numbers of recalled rules rose significantly, but the classification accuracy equal to or even higher than the reference points was detected quite soon in processing, for *InfoGain* for female dataset after selection of just four highest ranking attributes, for male writers and *MREVM* for just three most important features.

3.5 Pruning of Rule Sets Through Rule Rankings

Calculation of QM measure for rules can be understood as translating feature rankings into rule rankings. Depending on cardinalities of subsets of rules selected at each step, the total number of executed steps can significantly vary. The minimum is obviously one, while the maximum can even equal the total number of rules in the analysed set, if with each step only a single rule is added.

Table 2. Characteristics of decision algorithms with pruning of rules referring to specific conditional attributes: N indicates the number of considered attributes, (a) number of recalled rules, (b) maximal classification accuracy [%], (c) minimal support required of rules, (d) number of rules satisfying condition on support

N	Female								Male							
	InfoGain				MREVM				InfoGain				MREVM			
	(a)	(b)	(c)	(d)	(a)	(b)	(c)	(d)	(a)	(b)	(c)	(d)	(a)	(b)	(c)	(d)
1	10	61.11	55	4	10	61.11	55	10	6	13.33	14	4	6	13.33	14	4
2	27	81.11	55	13	27	81.11	55	13	15	21.11	9	8	27	55.56	9	23
3	36	81.11	55	13	56	82.22	32	27	45	61.11	6	35	80	**80.00**	25	39
4	79	**86.67**	55	27	97	81.11	55	14	73	61.11	10	41	127	**80.00**	25	50
5	91	86.67	55	27	203	82.22	44	26	153	75.56	21	62	219	**81.11**	20	115
6	128	86.67	55	27	324	**86.67**	55	28	198	75.56	21	65	290	80.00	**41**	25
7	167	86.67	55	27	578	86.67	55	30	239	75.56	26	46	562	75.56	41	28
8	202	86.67	55	27	877	86.67	**66**	13	307	75.56	21	72	778	75.56	41	29
9	356	86.67	**66**	11	1317	86.67	66	13	422	75.56	21	79	1073	75.56	41	30
10	570	86.67	66	11	1923	86.67	66	15	531	75.56	21	89	1355	75.56	41	31
11	1011	86.67	66	12	2755	86.67	66	16	689	75.56	32	44	1591	**78.89**	41	41
12	1415	86.67	66	14	3793	86.67	66	16	866	75.56	32	48	1975	**76.67**	41	45
13	2201	86.67	66	14	4995	86.67	66	**17**	1395	75.56	32	65	3169	76.67	41	45
14	3137	86.67	66	14	6671	86.67	66	17	1763	75.56	32	67	4456	**78.89**	41	53
15	4215	86.67	66	14	8099	86.67	66	17	2469	75.56	32	67	5774	**78.89**	41	53
16	5473	86.67	66	14	9485	86.67	66	17	3744	75.56	**41**	42	8476	76.67	41	63
17	7901	86.67	66	14	13255	86.67	66	17	4336	75.56	41	56	11055	76.67	41	66
18	10732	86.67	66	14	17589	86.67	66	17	5352	75.56	41	57	13428	76.67	41	69
19	14187	86.67	66	16	21238	86.67	66	17	7214	75.56	41	60	16188	76.67	41	69
20	18087	86.67	66	**17**	26821	86.67	66	17	9819	75.56	41	63	22035	76.67	41	75
21	23408	86.67	66	17	33834	86.67	66	17	14282	75.56	41	64	26674	76.67	41	78
22	31050	86.67	66	17	43225	86.67	66	17	18590	75.56	41	64	30846	76.67	41	78
23	39235	86.67	66	17	52587	86.67	66	17	26474	75.56	41	70	36630	76.67	41	78
24	48583	86.67	66	17	58097	86.67	66	17	35014	**76.67**	41	79	40024	76.67	41	78
25	62383	86.67	66	17					46191	76.67	41	**80**				

On the other hand, once the core sets of rules, corresponding to the decision algorithms limited by constraints on minimal support of rules and giving the best results for the complete algorithms, are retrieved, there is little point in continuing, thus the results presented in Table 3 stop when only fractions of the whole rule sets are recalled, for female writers just few hundreds, and for male writers close to ten thousand (still less than a quarter of the original algorithm).

3.6   Summary of the Best Results

Out of the two tested and compared approaches to rule filtering, selection governed by attributes included when following their rankings enabled to reject more rules from the reference algorithms, even over 35 % and 48 %, respectively for female and male datasets, with prediction at the reference level. For male writers recognition could be increased (at maximum by over 4 %) either with keeping or lowering constraints on minimal support required of rules.

Table 3. Characteristics of decision algorithms with pruning of rules while weighting them by measures based on rankings of conditional attributes: N indicates the weighting step, (a) number of recalled rules, (b) maximal classification accuracy [%], (c) minimal support required of rules, (d) number of rules satisfying condition on support

N	Female								Male							
	InfoGain-RDD				MREVM-RDD				InfoGain-RDD				MREVM-RDD			
	(a)	(b)	(c)	(d)	(a)	(b)	(c)	(d)	(a)	(b)	(c)	(d)	(a)	(b)	(c)	(d)
1	10	61.11	55	4	10	61.11	55	4	36	55.56	9	26	27	55.56	9	23
2	12	61.11	55	4	12	61.11	55	4	113	61.11	13	58	48	61.11	13	39
3	29	81.11	55	13	39	83.33	32	23	128	61.11	13	62	60	61.11	13	45
4	46	**87.78**	52	25	55	84.44	14	37	154	61.11	13	70	71	61.11	13	53
5	48	87.78	52	25	70	84.44	14	45	185	66.67	10	99	112	**80.00**	25	52
6	67	87.78	52	25	104	**87.78**	52	28	215	66.67	10	120	127	73.33	26	56
7	71	87.78	52	25	129	87.78	52	31	231	66.67	10	130	149	73.33	26	63
8	80	**90.00**	46	29	161	87.78	52	36	265	73.33	26	86	189	73.33	26	66
9	94	90.00	46	33	182	87.78	52	39	301	73.33	26	90	251	73.33	26	79
10	106	90.00	46	33	212	**88.89**	52	45	329	73.33	26	99	288	73.33	26	87
11	131	90.00	46	38	226	88.89	52	48	384	73.33	26	110	331	73.33	**41**	33
12	166	**86.67**	66	12	265	**86.67**	**66**	16	396	73.33	26	116	368	73.33	41	41
13	181	86.67	66	14	279	86.67	66	**17**	511	73.33	26	124	382	73.33	41	44
14	202	86.67	66	14	327	86.67	66	17	667	75.56	25	143	451	73.33	41	48
15	206	86.67	66	14	339	86.67	66	17	794	75.56	32	91	483	75.56	27	130
16	221	86.67	66	14	362	86.67	66	17	912	73.33	32	94	514	**76.67**	27	135
17	237	86.67	66	14	388	86.67	66	17	949	73.33	26	148	624	75.56	37	74
18	268	86.67	66	14	441	86.67	66	17	1011	73.33	**41**	54	848	75.56	37	77
19	285	86.67	66	16	452	86.67	66	17	1117	75.56	27	153	937	**78.89**	35	87
20	305	86.67	66	**17**	498	86.67	66	17	1189	75.56	27	155	1236	76.67	35	91
21									1228	75.56	27	157	1965	76.67	41	65
22									1900	75.56	41	61	2160	76.67	41	67
23									1993	75.56	41	63	2264	76.67	41	68
24									2667	**76.67**	41	67	3291	76.67	41	71
25									3610	76.67	41	68	4036	76.67	41	72
26									4577	76.67	41	70	4519	76.67	41	74
27									4825	76.67	41	71	5637	76.67	41	76
28									5725	76.67	41	74	6269	76.67	41	77
29									7901	76.67	41	76	9820	76.67	41	79
30									9250	76.67	41	78	9830	76.67	41	**80**
31									9394	76.67	41	79	9841	76.67	41	80
32									9404	76.67	41	**80**	9844	76.67	41	80

When rules were wighted, ranked, and then selected the quality of prediction was enhanced at maximum by over 3 % for both datasets, and for female and male writers datasets respectively over 29 % and 18 % of rules could be pruned.

For female dataset for both approaches to rule pruning better results were obtained while exploiting *InfoGain* attribute ranking, and for male dataset the same can be stated for *MREVM* ranking.

4   Conclusions

The paper presents research on selection of decision rules while following rankings of considered conditional attributes and exploiting weights assigned to them, which constitute alternatives to the popular approaches to rule filtering. Two ways to prune rules were compared, the first relying on selection of the rules with conditions only on the highest ranking attributes, while those referring to lower ranking features were rejected. Within the second methodology, the weights of attributes from their rankings formed a base from which for all rules the defined quality measures were calculated, and their values led to rule rankings. Next, the highest ranking rules were filtered out. For both described approaches two attribute rankings were tested, and the test results show several possibilities of constructing optimised rule classifiers, either with increased recognition, decreased lengths of decision algorithms, or both.

Acknowledgments. The research presented was performed at the Silesian University of Technology, Gliwice, Poland, within the project BK/RAu2/2016.

References

1. Amin, T., Chikalov, I., Moshkov, M., Zielosko, B.: Relationships between length and coverage of decision rules. Fundamenta Informaticae **129**, 1–13 (2014)
2. Baron, G.: On approaches to discretization of datasets used for evaluation of decision systems. In: Czarnowski, I., Caballero, A., Howlett, R., Jain, L. (eds.) Intelligent Decision Technologies 2016. Smart Innovation, Systems and Technologies, vol. 56, pp. 149–159. Springer, Switzerland (2016)
3. Bayardo Jr., R., Agrawal, R.: Mining the most interesting rules. In: Proceedings of the 5th ACM SIGKDD International Conference on Knowledge Discovery and Data Mining, pp. 145–154 (1999)
4. Cyran, K.A., Stanczyk, U.: Indiscernibility relation for continuous attributes: application in image recognition. In: Kryszkiewicz, M., Peters, J.F., Rybiński, H., Skowron, A. (eds.) RSEISP 2007. LNCS (LNAI), vol. 4585, pp. 726–735. Springer, Heidelberg (2007)
5. Fürnkranz, J., Gamberger, D., Lavrač, N.: Foundations of Rule Learning. Springer, Heidelberg (2012)

6. Gruca, A., Sikora, M.: Rule based functional description of genes – estimation of the multicriteria rule interestingness measure by the UTA method. Biocybernetics Biomed. Eng. **33**, 222–234 (2013)
7. Mansoori, E.: Using statistical measures for feature ranking. Int. J. Pattern Recog. Artitf. Intell. **27**(1), 1350003–1350014 (2013)
8. Sikora, M.: Induction and pruning of classification rules for prediction of micro-seismic hazards in coal mines. Expert Syst. Appl. **38**(2), 6748–6758 (2013)
9. Sikora, M., Wróbel, Ł.: Data-driven adaptive selection of rules quality measures for improving the rules induction algorithm. In: Kuznetsov, S.O., Ślęzak, D., Hepting, D.H., Mirkin, B.G. (eds.) RSFDGrC 2011. LNCS (LNAI), vol. 6743, pp. 278–285. Springer, Heidelberg (2011). doi:10.1007/978-3-642-21881-1_44
10. Słowiński, R., Greco, S., Matarazzo, B.: Dominance-based rough set approach to reasoning about ordinal data. In: Kryszkiewicz, M., Peters, J.F., Rybinski, H., Skowron, A. (eds.) RSEISP 2007. LNCS (LNAI), vol. 4585, pp. 5–11. Springer, Heidelberg (2007). doi:10.1007/978-3-540-73451-2_2
11. Stańczyk, U.: Decision rule length as a basis for evaluation of attribute relevance. J. Intell. Fuzzy Syst. **24**(3), 429–445 (2013)
12. Stańczyk, U.: Selection of decision rules based on attribute ranking. J. Intell. Fuzzy Syst. **29**(2), 899–915 (2015)
13. Stańczyk, U.: Measuring quality of decision rules through ranking of conditional attributes. In: Czarnowski, I., Caballero, A., Howlett, R., Jain, L. (eds.) Intelligent Decision Technologies 2016. Smart Innovation, Systems and Technologies, vol. 56, pp. 269–279. Springer, Switzerland (2016)

Stochastic Modelling

Energy Consumption Model for Data Processing and Transmission in Energy Harvesting Wireless Sensors

Yasin Murat Kadioglu[✉]

Intelligent Systems and Networks Group,
Department of Electrical and Electronic Engineering,
Imperial College, London SW7 2BT, UK
y.kadioglu14@imperial.ac.uk

Abstract. This paper studies energy harvesting wireless sensor nodes in which energy is gathered through harvesting process and data is gathered through sensing from the environment at random rates. These packets can be stored in node buffers as discrete packet forms which were previously introduced in "Energy Packet Network" paradigm. We consider a standby energy loss in the energy buffer (battery or capacitor) in a random rate, due to the fact that energy storages have self discharge characteristic. The wireless sensor node consumes K_e and K_t amount of harvested energy for node electronics (data sensing and processing operations) and wireless data transmission, respectively. Therefore, whenever a sensor node has less than K_e amount of energy, data can not be sensed and stored, and whenever there is more than K_e amount of energy, data is sensed and stored and also it could be transmitted immediately if the remaining energy is greater or equal than the K_t. We assume that the values of both K_e and K_t as one energy packet, which leads us a one-dimensional random walk modeling for the transmission system. We obtain stationary probability distribution as a product form solution and study on other quantities of interests. We also study on transmission errors among a set of M identical sensor with the presence of interference and noise.

Keywords: Wireless sensors · Energy harvesting · Energy packets · Data packets · Standby energy loss · Energy leakage · Data leakage · Markov modeling

1 Introduction

Wireless sensor network (WSN) is an essential part of IoT, which is composed of several sensors to sense physical data from the environment. The sensed data may be processed, stored, and transmitted by the sensor and communicate with a user or observer via Internet. A WSN can be used in many different areas such as [1]: health monitoring [2], environmental and earth sensing [3], industrial monitoring [4], and military applications [5]. Several application areas increase the usage of

© The Author(s) 2016
T. Czachórski et al. (Eds.): ISCIS 2016, CCIS 659, pp. 117–125, 2016.
DOI: 10.1007/978-3-319-47217-1_13

WSN numerously. While the worldwide number of the wireless-sensing points available is 4 million in 2011, more than 25 million available wireless-sensing points would be expected by 2017 [6], so that the envisaged market rise of WSN is from \$0.5 billion in 2012 to \$2 billion in 2022 [7].

When all energy is consumed in a sensor, it can not operate properly and can not achieve its role unless a new energy source is provided. However, replacing batteries or maintaining line connection for WSN usage is not convenient, so that the finite energy sources is a major constraint of WSNs. This has pushed to find an alternative energy source for WSNs, so that harvesting ambient energy from the environment has been addressed this problem and it has particular importance among these systems.

Earlier works [8,9] studied the performance of an energy harvesting sensor node as a function of random data and energy flow. Moreover, in [10,11] performance analysis was improved by taking into account the energy leakage from the storage due to standby operation, and [12] studied the case where exactly K energy packets are needed for successful transmission of 1 data packet. In earlier works, one of the main assumptions is that energy is only consumed for the packet transmission, not packet sensing and processing operations in the node. In this paper, the main contribution is that we consider energy consumption not only for data transmission but also for node electronics, i.e., data sensing-processing-stroring in an energy harvesting wireless sensor. The quantities of interest such as stationary probability distributions, excessive packet rates, and backlog probabilities for stability analysis is obtained. We also consider the transmission errors for the system and study on relation between system parameters and error probabilities.

2   Mathematical Model

We model a wireless sensor node where data and energy is received randomly from the environment. The arrivals of data packets and energy packets to the node are assumed to be independent Poisson process with rates λ and Λ, respectively. The term "energy packet" is a paradigm where energy is assumed to be in a discrete form. The sensor node contains a data buffer and an energy storage (capacitor or battery) to store receiving packets. Due to self discharge nature of energy storages, there is a standby loss in the system, that can be modeled as another independent Poisson process with rate μ. The sensing and the transmission occurs very fast at the node compared to the data and energy gathering rates from the environment, so that the operation times required for sensing and transmission processes are negligible, i.e., they occur instantaneously. In a sensor node, the harvested energy is basically consumed for packet sensing, storing, processing and transmission. In our system, we assume that $K_e = 1$ energy packet is required for the node electronics (sensing, storing, processing) and $K_t = 1$ energy packet is required for the data transmission, so that total two energy packets are needed for transmitting one data packet. Therefore, whenever a sensor node has less than $Ke = 1$ energy packet, data can not be sensed and

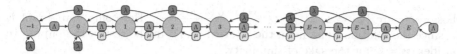

Fig. 1. State diagram representation of the system

stored, and whenever there is more than Ke amount of energy packet, data is sensed and stored and also it could be transmitted immediately if the remaining energy is greater or equal than the $Kt = 1$ energy packet.

Consider the system at a time $t \geq 0$ contains amount of $D(t)$ data packets in the buffer and amount of $E(t)$ energy packets in the storage, so that we can model the state of sensor node by the pair of $(D(t), E(t))$. Whenever $E(t) \geq 1$, node can sense the data packet and one energy packet is consumed by the node electronics instantaneously. Also, if there is still available energy in the storage, node can also transmit the data packet by consuming one more energy packet immediately.

When we examine the system model carefully, since the model has a finite state space, an unbounded growth of data or energy packets is not allowed. In fact, when one data packet arrives to the node whose state is $(D(t) = 0, E(t) = 1)$, the state will change as $(D(t) = 1, E(t) = 0)$ and it is the only state where data buffer is not empty. This interesting situation leads the system has great amount of excessive data packets, which we will consider later.

Let us write $p(d, e, t) = Prob[D(t) = d, \ E(t) = e]$. By using above remark, we should only consider $p(d, e, t)$ for the state space S such that $(e - d) \in S$, where $E \geq (e - d) \geq -1$ and E is the maximum amount of energy packets that can be stored in the node.

In fact, the system can be modeled as finite Markov chain whose states and transition diagram can be seen in Fig. 1. The stationary probabilities $p(e - d) = \lim_{t \to \infty} Prob[D(t) = d, \ E(t) = e]$ can be computed from following balance equations:

$$p(-1)[\Lambda] = \lambda \, p(1) \tag{1}$$

$$p(0)[\Lambda] = \Lambda \, p(-1) + \lambda \, p(2) + \mu \, p(1) \tag{2}$$

$$p(N)[\Lambda + \lambda + \mu] = \Lambda p(N - 1) + \lambda p(N + 2) + \mu p(N + 1) \tag{3}$$

$$p(E - 1)[\Lambda + \lambda + \mu] = \Lambda \, p(E - 2) + \mu \, p(E) \tag{4}$$

$$p(E)[\lambda + \mu] = \Lambda \, p(E - 1). \tag{5}$$

Note that (3) is valid for $0 < N < E - 1$ and has a solution of the form:

$$p(N) = c \, \varphi^N \tag{6}$$

where c is an arbitrary constant and φ can be computed from following characteristic equation:

$$\lambda \varphi^3 + \mu \varphi^2 - (\Lambda + \lambda + \mu)\varphi + \Lambda = 0 \tag{7}$$

whose roots are $\{\varphi_1 = 1, \ \varphi_{2,3} = \frac{-(\lambda+\mu)\mp\sqrt{(\lambda+\mu)^2+4\Lambda\lambda}}{2\lambda}$. Here only viable root is φ_3, since the solution must lie in the interval $(0,1)$. In the rest of the paper, we consider $\varphi_3 = \varphi$ for the sake of simplicity.

After finding stationary probabilities of the states between the interval $(0, E-1)$, we may also reach:

$$p(-1) = c\frac{\lambda}{\Lambda}\varphi, \quad p(0) = c(\frac{\lambda}{\Lambda}\varphi^2 + \frac{\lambda+\mu}{\Lambda}\varphi),$$

$$p(E-1) = c[1 + \frac{\lambda+\mu}{\Lambda} - \frac{\mu}{\lambda+\mu}]^{-1}\varphi^{E-2},$$

$$p(E) = c[(\frac{\lambda+\mu}{\Lambda})(\frac{\Lambda+\lambda+\mu}{\Lambda}) - \frac{\mu}{\Lambda}]^{-1}\varphi^{E-2}.$$

Using the fact that summation of the probabilities is one:

$$\sum_{N=-1}^{E} p(N) = c(\frac{2\lambda+\mu}{\Lambda}\varphi + \frac{\lambda}{\Lambda}\varphi^2) + c\sum_{N=1}^{E-2}\varphi^N + c[\frac{\lambda+\mu}{\Lambda} - \frac{\mu}{\Lambda+\lambda+\mu}]^{-1}\varphi^{E-2} = 1.$$

After further calculations, we may reach:

$$c = [\frac{2\lambda+\mu}{\Lambda}\varphi + \frac{\lambda}{\Lambda}\varphi^2 + \frac{\varphi - \varphi^{E-1}}{1 - \varphi} + \frac{\Lambda(\Lambda+\lambda+\mu)}{(\lambda+\mu)(\Lambda+\lambda+\mu) - \mu\Lambda}\varphi^{E-2}]^{-1}.$$

2.1   Excessive Packets Due to Finite Buffer Sizes

Since the energy storage capacity (maximum E energy packets) and data buffer capacity (maximum B data packets) are finite and data buffer is forced to be empty most of the time, we have some excessive packets that arrive at the node, but can not be sensed and stored. These excessive packets rates, Γ_d and Γ_e for data and energy packets, respectively and can be computed as:

$$\Gamma_d = \lambda\sum_{N=0}^{-B} p(N) = \lambda(p(0) + p(-1)) = c\lambda(\frac{2\lambda+\mu}{\Lambda}\varphi + \frac{\lambda}{\Lambda}\varphi^2),$$

$$\Gamma_e = \Lambda p(E) = c[\frac{1}{\Lambda}[(\frac{\lambda+\mu}{\Lambda})(\frac{\Lambda+\lambda+\mu}{\Lambda}) - \frac{\mu}{\Lambda}]]^{-1}\varphi^{E-2}.$$

Obviously, increase in the arrival rates of the energy and the data packets will increase the excessive packet rates. We can observe Γ_d remains zero until a certain level of λ and after this level, it starts showing a decreasingly growing behavior in Fig. 2, where we assume $\Lambda = 10, \mu = 1, E = 100$ for several values of λ. Although the system does not allow to store more than one data packet in the buffer, we observe reasonable amount of excessive data packet rate, which is due to the fact that most of the data packets can be sensed and transmitted when there are two or more energy packets in the node.

In Fig. 2, we may also observe similar effect on Γ_e where we assume $\lambda = 10, \mu = 0.1\Lambda, E = 100$ for several values of Λ. Apart from the previous observation, increase in Γ_e is nearly linear after a certain level of Λ.

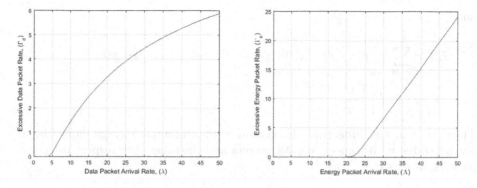

Fig. 2. Excessive data and energy packet rates.

2.2 Stability of the System

System stability is the question of whether finite number of segregated data packets and energy packets remain finite with certain probability for unlimited data and energy storage capacity when $t \to \infty$. If the condition is satisfied, then the system will be said to be stable.

Here, in order to make further analysis, we need to re-consider system with unlimited storages. In this case, we may reach:

$$p(-1) = c'\frac{\lambda}{\Lambda}\varphi, \ p(0) = c'(\frac{\lambda}{\Lambda}\varphi^2 + \frac{\lambda+\mu}{\Lambda}\varphi), \ p(N) = c'\varphi^N, \ 0 < N < \infty.$$

where φ is the same with the one solution of 7 and c' value can be computed as:

$$c' = \frac{(\lambda + \mu - 2\Lambda) + \sqrt{(\lambda+\mu)^2 + 4\Lambda\lambda}}{2(2\lambda + \mu)}.$$

Also, we can express the marginal probabilities as:

$$p_d(d) = \sum_{e=0}^{\infty} p(e - d) \text{ and } p_e(e) = \sum_{d=0}^{\infty} p(e - d).$$

In steady state, the probabilities that segregated data packets and energy packets do not exceed some finite values D' and E', respectively:

$$P_d(D') = lim_{t\to\infty} Prob[0 < D(t) \le D' < \infty], \tag{8}$$
$$P_e(E') = lim_{t\to\infty} Prob[0 < E(t) \le E' < \infty]. \tag{9}$$

We can calculate 8 and 9 by using marginal probabilities:

$$P_d(D') = \sum_{d=0}^{D'}\sum_{e=0}^{\infty} p(e - d) = p_d(1) + p_d(0) = p(-1) + p(0) + \sum_{N=1}^{\infty} c'\varphi^N = 1.$$

and

$$P_e(E') = \sum_{e=0}^{E'} \sum_{d=0}^{\infty} p(e-d) = p_e(0) + p_e(e)1[e > 0] = p(-1) + p(0) + \sum_{N=1}^{E'} c'\varphi^N$$
$$= 1 - c'\frac{\varphi^{E'+1}}{1-\varphi}.$$

Thus, we can conclude that the system with unlimited storage capacities is always stable with respect to data packets and unstable with respect to energy packets, as expected.

3   Analysis of Transmission Error Among a Set of Nodes

The total power that is entering the sensor node is simply energy harvesting rate Λ, due to the fact that energy rate is in unit of power. All harvested power can not be used by the node, since there are some energy packet losses, namely standby loss due to the self-discharge nature of the storage and excessive packet loss due to limited capacity storage of the node, so that the total power consumed by the node is:

$$\xi_i = \Lambda_i - \Gamma_{e_i} - \mu_i \sum_{N=1}^{E} p_i(N), \tag{10}$$

where the subscript i relates to the parameters of the i-th node among the set of M nodes. Whenever a node transmits a data packet, it consumes amount of K_e and K_t energy packets for node electronics and packet transmission, respectively. Since it is assumed that $K_e = K_t$, the total radiating power from a sensor on average is simply:

$$\phi_i = \frac{\xi_i}{2}. \tag{11}$$

Furthermore, if the probability of correctly receiving (or decoding) the packet sent by a given node i that transmits at power level K_{t_i} be denoted by:

$$1 - e_i = f(\frac{\eta_i K_{t_i}}{I_i + B_i}), \tag{12}$$

where f is some increasing function of its argument which is the signal to interference I_i plus noise B_i and $0 \leq \eta_i \leq 1$ represents the propagation factor of the transmission power that is sensed by the receiver.

Some number of 'α' separate frequency channels may be used in the communication medium. If the number of transmitting sensor nodes does not exceed α, distinct frequency channels are being used by each transmitter. In this case, interference can be considered as $I_{i_1} = \eta_i \kappa_{0_i}(M-1)\frac{\xi_i}{2}$, where $0 \leq \kappa_{0_i} \leq 1$ is a factor that represents the effect of side-band frequency channels and its value

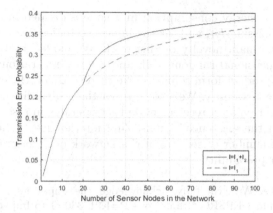

Fig. 3. Transmission error probability vs number of sensor nodes

is expected to be very small. On the other hand, if the number of transmitting sensor nodes exceeds α, some of the transmitters is forced to use a frequency channel already used by others, so that it will cause an additional interference $I_{i_2} = \kappa_i \frac{M-\alpha}{M} 1[M > \alpha]$, where κ_i is very close to 1 since interference is direct to the channel. Thus the total interference is:

$$I_i = I_{i_1} + I_{i_2} = \eta_i \frac{\xi_i}{2} \kappa_{0_i}(M-1) + \eta_i \frac{\xi_i}{2} (\frac{M-\alpha}{M}) 1[M > \alpha]. \qquad (13)$$

If we assume that all nodes are identical, we can replace (12) by:

$$1 - e = f(\frac{\eta K_t}{\eta \frac{\xi}{2} \kappa_0 (M-1) + \eta \frac{\xi}{2}(\frac{M-\alpha}{M}) 1[M > \alpha] + B}). \qquad (14)$$

Obviously, transmission error will raise with increase in number of sensor nodes in the network due to greater effect of the interference over the transmission. On the other hand, after a certain number of sensor nodes, α the system will face an additional interference, I_2 so that the error values will get higher values. We observe these effects in Fig. 3, where we assume that single bit transmission with $\Lambda = 10, \lambda = 10, \mu = 1, E = 100, B = 0.1, \eta = 0.5, \kappa_0 = 0.05, \alpha = 20$ and several values of M. Also, we assume BPSK transmission, so that:

$$1 - e = Q(\sqrt{\frac{\eta K_t}{\eta \frac{\xi}{2} \kappa_0 (M-1) + \eta \frac{\xi}{2}(\frac{M-\alpha}{M}) 1[M > \alpha] + B}}), \qquad (15)$$

where $Q(x) = \frac{1}{2}[1 - erf(\frac{x}{\sqrt{2}})]$.

4   Conclusions

This paper analyses wireless sensor nodes that gather both data and energy from the environment in random manners, so that they are able to operate autonomously. The energy consumption in a node is divided in two operations: for the data transmission K_t, and for the node electronics (sensing and processing) K_e that is the main novelty of this work. We modeled data transmission scheme as one-dimensional random walk and we express stationary probability distributions as a product form solution. We then study on the excessive packet rates and the system stability. We also consider the probability of a transmitted bit is correctly received by a receiver node that operates in a set of M identical sensor nodes with the existence of noise and interference. A numerical result show the effect of number of sensors in the network on interference values and transmission error probability.

Acknowledgments. We gratefully acknowledge the support of the ERA-NET ECROPS Project under EPSRC Grant No. EP=K017330=1 to Imperial College.

References

1. Yick, J., Mukherjee, B., Ghosal, D.: Wireless sensor network survey. Comput. Netw. **52**(12), 2292–2330 (2008)
2. Gao, T., Greenspan, D., Welsh, M., Juang, R., Alm, A.: Vital signs monitoring and patient tracking over a wireless network. In: 27th Annual International Conference of the Engineering in Medicine and Biology Society, IEEE-EMBS 2005, pp. 102–105, January 2005
3. Hart, J.K., Martinez, K.: Environmental sensor networks: a revolution in the earth system science? Earth Sci. Rev. **78**(3), 177–191 (2006)
4. Tiwari, A., Ballal, P., Lewis, F.L.: Energy-efficient wireless sensor network design and implementation for condition-based maintenance. ACM Trans. Sen. Netw. **3**, 1–23 (2007)
5. Yick, J., Mukherjee, B., Ghosal, D.: Analysis of a prediction-based mobility adaptive tracking algorithm. In: 2nd International Conference on Broadband Networks, BroadNets 2005, vol. 1, pp. 753–760, October 2005
6. Hatler, M.: Industrial wireless sensor networks: trends and developments. Retrieved **11**(14), 2013 (2013)
7. Harrop, P., Das, R.: Wireless sensor networks 2010–2020. Networks **2010**, 2020 (2010)
8. Gelenbe, E.: A sensor node with energy harvesting. ACM SIGMETRICS Perform. Eval. Rev. **42**(2), 37–39 (2014)

9. Gelenbe, E.: Synchronising energy harvesting and data packets in a wireless sensor. Energies **8**(1), 356–369 (2015)
10. Gelenbe, E., Kadioglu, Y.M.: Performance of an autonomous energy harvesting wireless sensor. In: Abdelrahman, O.H., Gelenbe, E., Gorbil, G., Lent, R. (eds.) Information Sciences and Systems 2015. LNEE, vol. 363, pp. 35–43. Springer, Heidelberg (2016). doi:10.1007/978-3-319-22635-4_3
11. Gelenbe, E., Kadioglu, Y.M.: Energy loss through standby and leakage in energy harvesting wireless sensors. In: 20th IEEE International Workshop on Computer Aided Modelling and Design of Communication Links and Networks (2015)
12. Kadioglu, Y.M., Gelenbe, E.: Packet transmission with K energy packets in an energy harvesting sensor. In: Proceedings of the 2nd International Workshop on Energy-Aware Simulation, ENERGY-SIM 2016, New York, NY, USA, pp. 1:1–1:6. ACM (2016)

Some Applications of Multiple Classes G-Networks with Restart

Jean Michel Fourneau[1]([✉]) and Katinka Wolter[2]

[1] DAVID, UVSQ, Versailles, France
jean-michel.fourneau@uvsq.fr
[2] Frei Universitat, Berlin, Germany

Abstract. We show how to model system management tasks such as load-balancing and delayed download with backoff penalty using G-networks with restart. We use G-networks with a restart signal, multiple classes or positive customers, PS discipline and arbitrary PH service distribution. The restart signal models the possibility to abort a task and send it again after changing its class and its service distribution. These networks have been proved to have a product form steady-state distribution.

Keywords: Performance · G-Networks · Phase-type distributions · Product form steady-state distribution · Restart

1 Introduction

Since the seminal papers [2,5,6] published by Gelenbe more than 20 years ago, G-networks of queues have received considerable attention. G-networks have been previously presented to model Random Neural Networks [7,8]. They contain queues, customers (like ordinary networks of queues) and signals which interact with the queues and disappear instantaneously. Due to these signals G-networks exhibit much more complex synchronization and allow to model new classes of systems (artificial or biological). Despite this complexity, most of the G-networks studied so far have a closed form solution for their steady-state.

For most of the results already known, the effect of the signal is the cancelation of customer or potential (for an artificial random neuron) [1]. Recently, we have studied G-networks with multiple classes where the signal is used to change the class of a customer in the queue [4]. Such a signal is denoted as a restart because in some models it is used to represent that a task is aborted and submitted again (i.e. restarted) when it encounters some problems (see [9,10] for some systems with restart). These models still have a product form steady-state solution under some technical conditions on the queue loads.

Here we present some examples to illustrate how this new model and theoretical result can help to evaluate the performance of a complex system. We hope that this result and the examples presented here open new avenues for research and applications of G-networks. The technical part of the paper is organized as follows. The model and the results proved in [4] are introduced in Sect. 2 while the examples are presented in Sect. 3.

T. Czachórski et al. (Eds.): ISCIS 2016, CCIS 659, pp. 126–133, 2016.
DOI: 10.1007/978-3-319-47217-1_14

2   Model Assumptions and Closed Form Solutions

We have considered in [4] generalized networks with an arbitrary number N of queues. We consider K classes of positive customers and only one class of signals. The external arrivals to the queues follow independent Poisson processes. The external arrival rate to queue i is denoted by $\lambda_i^{(k)}$ for positive customers of class k and Λ_i^- for signals. The customers are served according to the processor sharing (PS) policy. The service times are assumed to be Phase-type distributed, with one input (say 1) and one output state (say 0). At phase p, the intensity of service for customers of class k in queue i is denoted as $\mu_i^{(k,p)}$. The transition probability matrix $H_i^{(k)}$ describes how, at queue i, the phase of a customer of class k evolves. Thus the service in queue i is an excursion from state 1 to state 0 following matrix $H_i^{(k)}$ for a customer of class k. We consider a limited version of G-networks where the customers do not change into signals at the completion of a service. Here, customers may change class while they move between queues but they do not become signals. More precisely, a customer of class k at the completion of its service in queue i may join queue j as a customer of class l with probability $P_{i,j}^{+(k,l)}$. It may also leave the network with probability $d_i^{(k)}$. We assume that a customer cannot return to the queue it has just left: $P_{i,i}^{+(k,l)} = 0$ for all i, k and l. As usual, we have for all i,k: $\sum_{j=1}^N \sum_{l=1}^K P_{i,j}^{+(k,l)} + d_i^{(k)} = 1$.

Signals arrive from the outside according to a Poisson process of rate Λ_i^- at queue i. Signals do not stay in the network. Upon its arrival into a queue, a signal first choses a customer, then it interacts with the selected customer, and it finally vanishes instantaneously. If, upon its arrival, the queue is already empty, the signal also disappears instantaneously without any effect on the queue. The selection of the customer is performed according to a random distribution which mimics the PS scheduling. At state x_i, the probability for a customer to be selected is $\frac{x_i^{(k,p)}}{|x_i|} \mathbb{1}_{\{|x_i|>0\}}$ and the signal has an effect with probability $\alpha_i^{(k,p)}$. The effect is the restarting of the customer: this customer (remember it has class k and phase p) is routed as a customer of class l at phase 1 with probability $R_i^{(k,l)}$. We assume for all k, $R_i^{(k,k)} = 0$. Of course we have for all k, $\sum_{l=1}^K R_i^{(k,l)} = 1$ (Fig. 1).

The state of the queueing network is represented by the vector $x = (x_1, x_2, \ldots, x_N)$, where the component x_i denotes the state of queue i. As usual with multiple class PS queues with Markovian distribution of service, the state of queue i is given by the vector $(x_i^{(k,p)})$, for all class indices k and phase indices p. Clearly x is a Markov chain. Let us denote by $|x_i|$ the total number of customers in queue i. In [4] we have proved that the steady-state distribution, when it exists, has a product-form solution under some technical conditions on a fixed point system on the load.

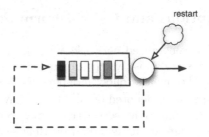

Fig. 1. Model of a queue with restart. The colors represent the classes

Theorem 1. *Consider an arbitrary open G-network with p classes of positive customers and a single class of negative customers the effect of which is to restart one customer in the queue. If the system of linear equations:*

$$\rho_i^{(k,1)} = \frac{\lambda_i^{(k)} + \sum_{o=1}^{P} \mu_i^{(k,o)} \rho_i^{(k,o)} H_i^{(k)}[o,1] + \nabla_i^{k,1} + \Delta_i^{k,1}}{\mu_i^{(k,1)} + \Lambda_i^- \alpha_i^{(k,1)}}, \tag{1}$$

where

$$\Delta_i^{k,1} = \sum_{p=1}^{P} \sum_{l=1}^{K} \Lambda_i^- \alpha_i^{(l,p)} \rho_i^{(l,p)} R_i^{(l,k)}, \tag{2}$$

$$\nabla_i^{k,1} = \sum_{j=1}^{N} \sum_{l=1}^{K} \sum_{q=1}^{P} \mu_j^{(l,q)} \rho_j^{(l,q)} H_j^{(l)}[q,0] P_{j,i}^{+(l,k)}, \tag{3}$$

and,

$$\forall p > 1, \quad \rho_i^{(k,p)} = \frac{\sum_{o=1}^{P} \mu_i^{(k,o)} \rho_i^{(k,o)} H_i^{(k)}[o,p]}{\mu_i^{(k,p)} + \Lambda_i^- \alpha_i^{(k,p)}} \tag{4}$$

has a positive solution such that for all stations i $\sum_{k=1}^{K} \sum_{p=1}^{P} \rho_i^{(k,p)} < 1$, then the system stationary distribution exists and has product form:

$$p(\boldsymbol{x}) = \prod_{i=1}^{N} (1 - \sum_{k=1}^{K} \sum_{p=1}^{P} \rho_i^{(k,p)}) |\boldsymbol{x}_i|! \prod_{k=1}^{K} \prod_{p=1}^{P} \frac{(\rho_i^{(k,p)})^{x_i^{(k,p)}}}{x_i^{(k,p)}!}. \tag{5}$$

Property 1. *This result is used to obtain closed form solutions for some performance measures: the probability to have exactly m customers in the queue and the expected number of customers in the queue.*

$$Pr(m \; customers) = (1 - \sum_{k=1}^{K}\sum_{p=1}^{P} \rho_i^{(k,p)}) \left[\sum_{k=1}^{K}\sum_{p=1}^{P} \rho_i^{(k,p)}\right]^m,$$

$$E[N] = \frac{\sum_{k=1}^{K}\sum_{p=1}^{P} \rho_i^{(k,p)}}{1 - \sum_{k=1}^{K}\sum_{p=1}^{P} \rho_i^{(k,p)}}. \tag{6}$$

3 Examples

We now present some examples to put more emphasis on the modeling capabilities of G-networks with restart signals. We model a load balancing system where the restarts are used to migrate the customers between queues and a back off mechanism for delayed downloading.

Example 1. Load Balancing: We consider two queues in parallel as depicted in Fig. 2. We want to represent a load balancing mechanism between them and we want to get the optimal rates to operate this mechanism and obtain the best performance.

The queues receive two types of customers: type 1 customers need to be served while type 2 customers represent the customers which must be moved to the other queue to balance the load. Customers of type 1 arrive from the outside according to two independent Poisson process with rate $\lambda_1^{(1)}$ for queue 1 and $\lambda_2^{(1)}$ for queue 2. There are no arrivals from the outside for type 2 customers. Type 2 customers are created by a restart. The service rates do not depend on the queue. They are equal to $\mu^{(1)}$ for type 1 and $\mu^{(2)}$ for type 2. For the sake of simplicity, we assume here that the service distributions are exponential. PH distributions will be added at the end of this example.

Restarting signals arrive to queues 1 and 2 according to two independent Poisson processes with rate Λ_1^- and Λ_2^-. When it arrives to a queue, a signal choses a customer at random as mentioned in the previous section and tries to change it to type 2. We assume the following probabilities of success: $\alpha_1^{(1)} = 1$ and $\alpha_1^{(2)} = 0$. Similarly, $\alpha_2^{(1)} = 1$ and $\alpha_2^{(2)} = 0$. Note that we have simplified the notation as we only have one phase of service (we consider exponential rather than PH distributions). This value of the acceptance probability means that the restarting signals is always accepted when the signal selects a type 1 customer and it fails when it tries to restart a type 2 customer (as by definition in this model, a type 2 customer is already restarted).

After its service, a type 1 customer leaves the system while a type 2 customer moves to the other queue and changes its type during the movement to become a type 1 customer. Thus the load balancing mechanism proceeds as follows: the signal is received by the queue and it selects a customer at random. If the customer has type 2, nothing happens. If the selected customer has type 1, it is restarted as a type 2 customer with another service time distribution and another routing matrix. The service time for a type 2 customer represents the

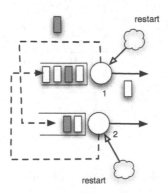

Fig. 2. Two queues in parallel with load balancing performed by restart signals

time needed to organize the job migration. It is assumed that it is much shorter than the the service type of a type 1 customer which represents the effective service. Let us now write the flow equations:

$$\rho_1^{(1)} = \frac{\lambda_1^{(1)} + \rho_2^{(2)}\mu^{(2)}}{\mu^{(1)} + \Lambda_1^-}, \quad \rho_1^{(2)} = \frac{\Lambda_1^- \rho_1^{(1)}}{\mu^{(2)}}, \quad \rho_2^{(1)} = \frac{\lambda_2^{(1)} + \rho_1^{(2)}\mu^{(2)}}{\mu^{(1)} + \Lambda_2^-}, \quad \rho_2^{(2)} = \frac{\Lambda_2^- \rho_2^{(1)}}{\mu^{(2)}}.$$

(7)

Let us now consider the performance of such a system. We control the system with the rate of arrival of signals Λ_1^- and Λ_2^- and the objective is to balance the load with the smallest overhead. More formally, we say that the system is balanced if the loads for customers in service (i.e. not preparing their migration) are equal for both queues (i.e. $\rho_1^{(1)} = \rho_2^{(1)} = \rho$) and we assume that the overhead is the load of the queues due to the migration (i.e. $\rho_1^{(2)} + \rho_2^{(2)}$). Assuming that the system is balanced, we have:

$$\rho = \frac{\lambda_1^{(1)} + \rho_2^{(2)}\mu^{(2)}}{\mu^{(1)} + \Lambda_1^-} = \frac{\lambda_2^{(1)} + \rho_1^{(2)}\mu^{(2)}}{\mu^{(1)} + \Lambda_2^-}$$

After substitution, we get: $\rho = \frac{\lambda_1^{(1)} + \rho\Lambda_2^-}{\mu^{(1)} + \Lambda_1^-} = \frac{\lambda_2^{(1)} + \rho\Lambda_1^-}{\mu^{(1)} + \Lambda_2^-}$. Without loss of generality we assume that $\lambda_1^{(1)} > \lambda_2^{(1)}$. Taking into account the first part of the equation, we obtain: $\rho(\Lambda_1^- - \Lambda_2^-) = \lambda_1^{(1)} - \rho\mu^{(1)}$. Similarly using the second equation we get:

$$\rho(\Lambda_1^- - \Lambda_2^-) = \rho\mu^{(1)} + \lambda_2^{(1)}.$$

Thus, $\rho = \frac{\lambda_1^{(1)} - \lambda_2^{(1)}}{2\mu^{(1)}}$, and $\Lambda_1^- - \Lambda_2^- = \frac{\lambda_1^{(1)} + \lambda_2^{(1)}}{2}$. Taking now the other part of the objective into account we want to minimize the overhead of the load balancing mechanism. Remember that the global overhead is:

$$\rho_1^{(2)} + \rho_2^{(2)} = \rho\frac{(\Lambda_1^- + \Lambda_2^-)}{\mu^{(2)}}.$$

Thus the optimal solution is achieved for $\Lambda_2^- = 0$ and $\Lambda_1^- = \frac{\lambda^{(1)} + \lambda_2^{(1)}}{2}$. Let us now consider a more complex problem where the services for type 1 customer follow the same PH distribution. We still assume that type 2 customers receive services with an exponential distribution. Let us now write the flow equations:

$$\rho_1^{(1,1)} = \frac{\lambda_1^{(1)} + \sum_{p>0} \rho_2^{(2,p)} \mu^{(2,p)}}{\mu^{(1,1)} + \Lambda_1^-}, \qquad \rho_1^{(1,p)} = \frac{H(1,p)\rho_1^{(1,1)} \mu^{(1,1)}}{\mu^{(1,p)} + \Lambda_1^-}, \forall p > 1,$$

$$\rho_2^{(1,1)} = \frac{\lambda_2^{(1)} + \sum_{p>0} \rho_1^{(2,p)} \mu^{(2,p)}}{\mu^{(2,1)} + \Lambda_2^-}, \qquad \rho_2^{(1,p)} = \frac{H(1,p)\rho_2^{(2,1)} \mu^{(2,1)}}{\mu^{(2,p)} + \Lambda_2^-}, \forall p > 1, \qquad (8)$$

$$\rho_1^{(2)} = \frac{\Lambda_1^- \sum_{p>0} \rho_1^{(1,p)}}{\mu^{(2)}}, \qquad \rho_2^{(2)} = \frac{\Lambda_2^- \sum_{p>0} \rho_2^{(1,p)}}{\mu^{(2)}}.$$

These equations can be used to optimize the system as we have done previously for exponential service distributions.

Example 2. Delayed Downloading: We now study a small wifi network with a delayed downloading mechanism (see for instance [11]). Queue A is the downloading queue (see Fig. 3). Customers and signals arrive from the outside to queue A. The class of customers represents the delays that requests will experience. Type 1 requests (in white) are not delayed while delayed requests are depicted in grey. The restart signals change the state of a request to "delayed" according to the selection mechanism described in Sect. 2. The probability of acceptance for the selection depends on the class of the customer and the phase of service. Thus, we can model delay based on the steps of the downloading protocol, for instance. Once a request class has been changed due to selection by the signal, it is routed after its service to queue B or C where it is changed again to a class 1 request and experiences a random delay depending on the queue. The flow equations are:

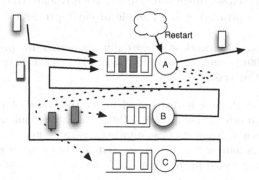

Fig. 3. The queuing network associated to the delayed downloading with back-off penalties

$$\rho_A^{1,1} = \frac{\sum_{o=1}^{P} \mu_A^{1,o} \rho_A^{1,o} H_A^{(k)}[o,1] \; + \; \sum_{p=1}^{P} \mu_B^{1,p} \rho_B^{1,p} H_B^{(1)}[p,0] \; + \; \sum_{p=1}^{P} \mu_C^{1,p} \rho_C^{1,p} H_C^{(1)}[p,0]}{\mu_A^{1,1} + \Lambda_A^- \alpha_A^{1,1}}, \tag{9}$$

$$\forall p > 1, \quad \rho_A^{1,p} = \frac{\sum_{o=1}^{P} \mu_A^{1,o} \rho_A^{1,o} H_A^{(1)}[o,p]}{\mu_A^{1,p} + \Lambda_A^- \alpha_A^{1,p}}, \quad and \;\; \forall k > 1, \rho_A^{k,p} = \frac{\sum_{o=1}^{P} \mu_A^{k,o} \rho_A^{k,o} H_A^{(k)}[o,p]}{\mu_A^{k,p}}, \tag{10}$$

$$\rho_A^{k,1} = \frac{\sum_{o=1}^{P} \mu_A^{k,o} \rho_A^{k,o} H_A^{(k)}[o,1] \; + \; \sum_{p=1}^{P} \Lambda_A^- \alpha_A^{(1,p)} \rho_A^{(1,p)} R_A^{(1,k)}}{\mu_A^{1,1} + \Lambda_A^- \alpha_A^{1,1}}, \tag{11}$$

$$\rho_B^{1,1} = \frac{\sum_{o=1}^{P} \mu_B^{1,o} \rho_B^{1,o} H_B^{(k)}[o,p] \; + \; \sum_{p=1}^{P} \mu_A^{2,p} \rho_A^{2,p} H_A^{(2)}[p,0]}{\mu_B^{1,1}}, \tag{12}$$

$$\forall p > 1, \quad \rho_B^{1,p} = \frac{\sum_{o=1}^{P} \mu_B^{1,o} \rho_B^{1,o} H_B^{(k)}[o,1]}{\mu_B^{1,p}}, \quad and \;\; \rho_C^{1,p} = \frac{\sum_{o=1}^{P} \mu_C^{1,o} \rho_C^{1,o} H_C^{(k)}[o,1]}{\mu_C^{1,p}}, \tag{13}$$

$$\rho_C^{1,1} = \frac{\sum_{o=1}^{P} \mu_C^{1,o} \rho_C^{1,o} H_C^{(k)}[o,p] \; + \; \sum_{p=1}^{P} \mu_A^{3,p} \rho_A^{3,p} H_A^{(3)}[p,0]}{\mu_C^{1,1}}. \tag{14}$$

Assuming that these equations have a fixed point solution such that the queues are stable, Theorem 1 proves that the steady-state distribution has product form. This closed form solution allows us to study the performance of the downloading mechanism and to optimize the throughput when one changes the delay distributions.

4 Concluding Remarks

Note that it is possible to add triggers in the model to increase the flexibility while conserving the closed form solution [3]. We advocate that G-networks with restart signals are a promising and flexible modeling technique.

Acknowledgments. This work was partially supported by project MARMOTE (ANR-12-MONU-00019) and by a PROCOPE PHC grant between Université de Versailles and Frei Universitat, Berlin.

References

1. Artalejo, J.R.: G-networks: a versatile approach for work removal in queuing networks. European J. Op. Res. **126**, 233–249 (2000)
2. Fourneau, J.M., Gelenbe, E., Suros, R.: G-networks with multiple classes of positive and negative customers. Theor. Comput. Sci. **155**, 141–156 (1996)
3. Fourneau, J.-M., Wolter, K.: Mixed networks with multiple classes of customers and restart. In: Remke, A., Manini, D., Gribaudo, M. (eds.) ASMTA 2015. LNCS, vol. 9081, pp. 73–86. Springer, Heidelberg (2015)
4. Fourneau, J.M., Wolter, K., Reinecke, P., Krauß, T., Danilkina, A.: Multiple class G-networks with restart. In: ACM/SPEC International Conference on Performance Engineering, ICPE 2013, pp. 39–50. ACM (2013)
5. Gelenbe, E.: Product-form queuing networks with negative and positive customers. J. Appl. Probab. **28**, 656–663 (1991)
6. Gelenbe, E.: G-networks with instantaneous customer movement. J. Appl. Probab. **30**(3), 742–748 (1993)
7. Gelenbe, E.: G-networks: an unifying model for queuing networks and neural networks. Ann. Oper. Res. **48**(1–4), 433–461 (1994)
8. Gelenbe, E., Fourneau, J.M.: Random neural networks with multiple classes of signals. Neural Comput. **11**(4), 953–963 (1999)
9. van Moorsel, A.P.A., Wolter, K.: Analysis and algorithms for restart. In: 1st International Conference on Quantitative Evaluation of Systems (QEST 2004), The Netherlands, pp. 195–204. IEEE Computer Society (2004)
10. van Moorsel, A.P.A., Wolter, K.: Analysis of restart mechanisms in software systems. IEEE Trans. Softw. Eng. **32**(8), 547–558 (2006)
11. Wu, H., Wolter, K.: Analysis of the energy-performance tradeoff for delayed mobile offloading. In: Proceedings of the 9th EAI International Conference on Performance Evaluation Methodologies and Tools, VALUETOOLS 2015, pp. 250–258 (2015)

XBorne 2016: A Brief Introduction

Jean Michel Fourneau$^{(\boxtimes)}$, Youssef Ait El Mahjoub, Franck Quessette, and Dimitris Vekris

DAVID, UVSQ, Versailles, France
jean-michel.fourneau@uvsq.fr

Abstract. We present the new version of XBorne a software tool for the probabilistic modeling with Markov chains. The tool which has been developed initially as a testbed for the algorithmic stochastic comparisons of stochastic matrices and Markov chains, is now a general purpose framework which can be used for the Markovian modelling in education and research.

Keywords: Performance · Numerical analysis · Simulation · Markov chains

1 Introduction

The numerical analysis of Markov chains always deals with a tradeoff between complexity and accuracy. Therefore we need tools to compare the approaches, the codes and some well-defined examples to use as a testbed. After many years of development of exact or bounding algorithms for stochastic matrices, we have gathered the most efficient into XBorne, our numerical analysis tool [8]. Typically using XBorne, one can easily build models with tens of millions of states. Note that solving any questions with this size of models is a challenging issue. XBorne was developed with the following key ideas:

1. Build one software tool dedicated to only one function and let the tools communicate with file sharing
2. If another tool already exists for free and is sufficiently efficient, use it and write the export tool (only create tools you cannot find easily).
3. Allow to recompile the code to include new models.
4. Separate the data and the description of the data.

As a consequence, we have chosen to avoid the creation of a new modelling language. The models are written in C and included as a set of 4 functions to be compiled by the model generator. This aspect of the tool will be emphasized in Sect. 2 with the presentation of an example (a queue with hysteresis). The tool decomposition approach will also be illustrated in the paper.

XBorne is now a part of the French project MARMOTE which aims to build a set of tools for the analysis of Markovian models. It is based on PSI3 to perform perfect simulation (i.e. Coupling from the past) of monotone systems and

© The Author(s) 2016
T. Czachórski et al. (Eds.): ISCIS 2016, CCIS 659, pp. 134–141, 2016.
DOI: 10.1007/978-3-319-47217-1_15

their generalizations [5], MarmoteCore to provide an object interface to Markov objects and associated methods, and XBorne that we will present in this paper. The aim of XBorne (and the other tools developed in the MARMOTE project) is not to replace older modeling tools but to be included into a larger framework where we can share tools and models developed in well-specified frameworks which can be translated into one another. XBorne will be freely available upon request.

The technical part of the paper is as follows: in Sect. 2, we present how we can build a new model. We show in Sect. 3 how it can be solved and we present some numerical results. Sections 4 and 5 are devoted to two new solving techniques. In Sect. 4, we consider the quasi-lumpability technique. We modify the Tarjan and Paige approach used for the detection of macro-states for aggregation or bisimulation [12] to relax the assumption on the creation of macro states and accommodate a quasi-lumpable partition of the state space. Section 5 is devoted to the simulation of Markov chains and it is presented here to show how we have chosen to connect XBorne with other tools.

2   Building a Model with XBorne

XBorne can be used to generate a sparse matrix representation of a Discrete Time Markov Chain (DTMC) from a high level description provided in C. Continuous-time models can be considered after uniformization (see the example in the following). Like many other tools, the formalism used by XBorne is based on the description of the states and the transitions. All the information concerning the states and the transitions are provided by the modeler using 2 files (1 for the constants and one for the code, respectively denoted as "const.h" and "fun.c"). States belong to a hyper-rectangle the dimension of which is given by the constant NEt. The bounds of the hyper-rectangle must be given by function "InitEtendue()". The states belong to the hyper-rectangle and they are found by a BFS visit from an initial state given by the modeler through function "EtatInitial()".

The transitions are given in a similar manner. The constant "NbEvtsPossibles" is the number of events which provoke a transition. The idea is that an event is a mapping applied to a state (not necessarily a one to one mapping). Each event has a probability given by function "Probabilite()" and its value may depend on the state description. The mapping realized by an event is described by function "Equation()". To conclude, it is sufficient to describe 4 functions in C and some definitions and recompile the model generator to obtain a new code which builds the transition probability matrix.

```
#define NEt      2          #define NbEvtsPossibles  4
#define AlwaysOn  10         #define BufferSize  20
#define OnAndOff   5         #define UPandDOWN   0
#define WARMING    1         #define ALL_UP   2
#define UP        10         #define DOWN  5
```

We now present an example for the various definitions and functions which are written in the files "const.h" and "fun.c" to describe the model developed by Mitrani in [11] to study the tradeoff between energy consumption and quality of service in a data-center. It is a model of a M/M/(a+b) queue with hysteresis and impatience. We have slightly changed the assumptions as follows: the queue is finite with size "BufferSize". The arrivals still follow a Poisson process with rate "Lambda". The services are exponential with rate "Mu". Initially only "AlwaysOn" servers are available. Once the number of customers in the queue is larger than "UP", another set (with size OnAndOff) of servers is switched on. The switching time has an exponential duration with rate "Nu". If the number of customers becomes smaller than "DOWN", this set of servers is switched off. This action is immediate. As NEt=2, a state is a two dimension vector. The first dimension is the number of customers and the second dimension encodes the state of the servers. The initial state is an empty queue with the extra block of servers which is not activated.

```
void InitEtendue()
{
    Min[0] = 0;  Max[0] = BufferSize; Min[1] = UPandDOWN;  Max[1] = ALL_UP;
}
void EtatInitial(E)
int *E;
{
    E[0] = 0; E[1] = UPandDOWN;
}
double Probabilite(int indexevt, int *E) {
    double p1, Delta;
     int nbServer, inserv;
    nbServer = AlwaysOn;
    if (E[1]==ALL_UP) {nbServer += OnAndOff;}
    inserv = min(E[0], nbServer);
    Delta = Lambda + Nu + Mu*(AlwaysOn + OnAndOff);
    switch (indexevt) {
        case ARRIVAL: p1 = Lambda/Delta; break;
        case SERVICE: p1 = (inserv)*Mu/Delta; break;
        case SWITCHINGON: p1 = Nu/Delta; break;
        case LOOP: p1 = Mu*(AlwaysOn + OnAndOff - inserv)/Delta; break;
    }
    return(p1);
}
```

The model is in continuous time. Thus we build an uniformized version of the model adding a new event to generate the loops in the transition graph which are created during the uniformization. After this process we have 4 events: ARRIVAL, SERVICE, SWITCHINGON, LOOP. In all the functions, E and F are states. The generation tool creates 3 files: one contains the transition matrix in sparse row format, the second gives information on the number of states and transitions and the third one stores the encoding of the states. Indeed the states are found during the BFS visit of the graph and they are ordered by this visit algorithm. Thus, we have to store in a file the mapping between the state number given by the algorithm and the state description needed by the modeler and some algorithms.

```
void Equation(int *E, int indexevt, int *F, int *R)
{
    F[0] = E[0]; F[1] = E[1];
    switch (indexevt) {
        case ARRIVAL: if (E[0]<BufferSize) {F[0]++;}
                if ((E[0]>=UP) && (E[1]==UPandDOWN)) {F[1]=WARMING;}
                break;
        case SERVICE: if (E[0]>0) {F[0]--;}
                if ((F[0]==DOWN) && (E[1]>UpandDOWN)) {F[1]=UPandDOWN;}
                break;
        case SWITCHINGON: if (E[1]==WARMING) {F[1]=ALL_UP;}
                break;
        case LOOP: break;
    }
}
```

Once the steady-state distribution is obtained with some numerical algorithms, the marginal distributions and some rewards are computed using the description of the states obtained by the generation method and codes provided (and compiled) by the modeler to specify the rewards (see in the left part of Fig. 1 the marginal distribution for the queue size).

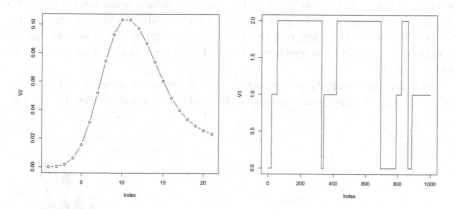

Fig. 1. Mitrani's model. Steady-state for the queue size (left). Sample path of the state of the servers (right).

3   Numerical Resolution

In XBorne, we have developed some well-known numerical algorithms to compute the steady-state distribution (GTH for small matrices), SOR and Gauss Seidel for large sparse matrices but we have chosen to export the matrices into

MatrixMarket format to use state of the art solvers which are now available on the web. But we also provide new algorithms for the stochastic bounds or the element-wise bound of the matrices, the stochastic bound or the entry-wise bounds of the steady-state distribution. These bounds are based on the algorithmic stochastic comparison of Discrete Time Markov Chain (see [10] for a survey) where stochastic comparison relations are mitigated with structural constraints on the bounding chains. More precisely, the following methods are available:

- Lumpability: to enforce the bounding matrix to be ordinary lumpable. Thus, we can aggregate the chain [9].
- Pattern based: to enforce the bounding matrix to follow a pattern which provides an ad-hoc numerical algorithm (think at a upper Hessenberg matrix for instance) [2].
- Censored Markov chain: only the useful part of the chain is censored and we provide bounds based on this partial representation of the chain [1,7].

Other techniques for entry-wise bounds of the steady state distribution have also been derived and implemented [3]. They allow in some particular cases to deal with infinite state space (otherwise not considered in XBorne).

More recently, we have developed a new low rank decomposition for a stochastic matrix [4]. This decomposition is adapted to stochastic matrices because it provides an approximation which is still a stochastic matrix while singular value decomposition gives a low rank matrix which is not stochastic anymore. Our low rank decomposition allows to compute the steady-state distribution and the transient distribution with a lower complexity which takes into account the matrix rank. For instance, for a matrix of rank k and size N, the computation of the steady-state distribution requires $O(Nk^2)$ operations. We also have derived algorithms to provide stochastic bounds with a given rank for any stochastic matrix (see [4]).

Note that the integration with other tools we mention previously is not limited to numerical algorithms provided by statistical package like R. We also

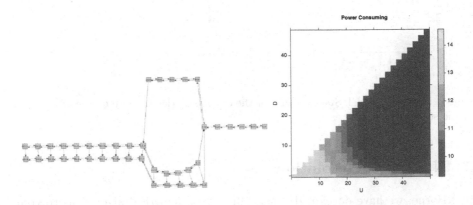

Fig. 2. Mitrani's model. Directed graph of the chain (left). Energy consumption (right).

use their graphic capabilities and the layout algorithms. We illustrate these two aspects in Fig. 2. In the left part we have drawn the layout of the Markov chain associated with Mitrani's model for a small buffer size (i.e. 20). We have developed a tool which reads the Markov chains description and write it as a labelled directed graph in "tgf" format. With this graph description, we use the graph editors available on the web to obtain a layout of the chain and to visualize the states and their transitions. On the right part of the figure, we have depicted a heat diagram for the energy consumption associated to Mitrani's model for all the values of the thresholds U and D.

4 Quasi-Lumpability

Quasi-Lumpability testing has been recently added into XBorne to analyze very large matrices. The numerical algorithms which have been developed are also used to analyze stochastic matrices which are not completely specified. It is well-known now that Tarjan's algorithm can be used to obtain the coarsest partition of the state space of a Markov chain which is ordinary lumpable and which is consistent with an initial partition provided by the modeler. Lumpable matrix can be aggregated to obtain a smaller matrix, easier to analyze. Logarithmic reduction in size are often reported in the literature. We define quasi-lumpability of partition $A_1, A_2, \ldots, A_k$ with threshold ϵ of stochastic matrix M as follows: for all macro-states A_i and A_j we have

$$\max_{l1, l2 \in A_i} \left| \sum_{k \in A_j} M(l1, k) - \sum_{k \in A_j} M(l2, k) \right| = E(i, j) \leq \epsilon. \tag{1}$$

When $\epsilon = 0$ we obtain the definition of ordinary lumpability. We have modified Tarjan's algorithm to obtain a partition which is quasi-lumpable given an initial partition and a maximum threshold ϵ. The output of the algorithm is the coarsest partition consistent with the initial partition and the real threshold needed in the algorithm (which can be smaller than ϵ). Note that the algorithm always returns a partition. However the partition may be useless as it may have a large number of nodes. The next step is to lump matrix M according to the partition found by the modified Tarjan's algorithm. If the real threshold needed is equal to 0, the matrix is lumpable and the aggregated matrix is stochastic. It is solved with classical methods.

If the threshold needed is positive, we obtain two aggregated matrices Up and Lo: one where the transition probability between macro states A_i and A_j is equal to $\max_{l \in A_i} \sum_{k \in A_j} M(l, k)$ and one where it is equal to $\min_{l \in A_i} \sum_{k \in A_j} M(l, k)$. Up is super-stochastic while Lo is sub-stochastic. These two bounding matrices also appear when the Markov chains are not completely specified and transitions are associated with intervals of probability. We have implemented Courtois and Semal algorithm [6] to obtain entry-wise bounds on the steady-state distribution of all matrices between Up and Lo. We are still conducting new research to improve this algorithm.

5   Simulation

We have added several simulation engines in XBorne, mainly for educational purpose and for verification. All of them define a model with the same functions we have previously presented to design a Markov chain. The modeler just needs to add the simulation time and the seed for the generator when a random number generator is used by the simulation code. Thus, the same model description (i.e. the four C functions) is used for the simulation and the Markov chain generation.

Two types of engines have been developed: a simulator with random number generation in C and a trace base version where the random number generation (and generally the random variables generation) are outside the simulation code and previously stored in a file by some statistical packages (typically R). Similarly, the output of the simulations are sample paths which are stored in separate files to be analyzed by state of the art statistical packages where various test algorithms and confidence intervals computations are performed by efficient methods already available in these packages. Thus, the modeler is expected to concentrate on the development of the model simulation, leaving the statistical details to other packages. Similarly, the drawing of the paths can be obtained from the statistical package like in the right part of Fig. 1 where we depict the evolution of the second component of Mitrani's model (i.e. the state of the server). The trace based simulation is also used to simulate Semi-Markov processes.

The simulation engines also differ by the definition of paths: the general purpose simulation engine builds one path per seed for the simulation time, while the regenerative Markovian simulation stores one path per regenerative cycle. Furthermore, to deal with the complexity of the simulation of discrete distribution by the reverse transform method, we have implemented two types of engine: a general inverse distribution method when the distribution of probability for the next event changes with the state, and an alias method when this distribution is the same for all the states.

Acknowledgments. This work was partially supported by project MARMOTE (ANR-12-MONU-00019). Y. Ait El Mahjoub is supported by Labex DigiCosme (project ANR-11-LABEX-0045-DIGICOSME) operated by ANR as part of the program Investissement d'Avenir Idex Paris-Saclay (ANR-11-IDEX-0003-02).

References

1. Busic, A., Djafri, H., Fourneau, J.M.: Bounded state space truncation and censored Markov chains. In: 51st IEEE Conference on Decision and Control (CDC 2012) (2012)
2. Busic, A., Fourneau, J.M.: A matrix pattern compliant strong stochastic bound. In: 2005 IEEE/IPSJ International Symposium on Applications and the Internet Workshops (SAINT Workshops), Italy, pp. 260–263. IEEE Computer Society (2005)
3. Busic, A., Fourneau, J.M.: Iterative component-wise bounds for the steady-state distribution of a Markov chain. Numer. Linear Algebra Appl. **18**(6), 1031–1049 (2011)
4. Busic, A., Fourneau, J.M., Ben Mamoun, M.: Stochastic bounds with a low rank decomposition. Stochast. Models **30**(4), 494–520 (2014). Special Issue with selected papers from the Eighth Int. Conf. on Matrix-Analytic Methods in Stochastic Models
5. Busic, A., Gaujal, B., Gorgo, G., Vincent, J.M.: Psi2: Envelope perfect sampling of non monotone systems. In: QEST 2010, Seventh International Conference on the Quantitative Evaluation of Systems, Virginia, USA, pp. 83–84. IEEE Computer Society (2010)
6. Courtois, P.J., Semal, P.: On polyhedra of Perron-Frobenius eigenvectors. Linear Algebra Appl. **65**, 157–170 (1985)
7. Dayar, T., Pekergin, N., Younès, S.: Conditional steady-state bounds for a subset of states in Markov chains. In: Structured Markov Chain (SMCTools) workshop in VALUETOOLS. ACM (2006)
8. Fourneau, J.M., Le Coz, M., Pekergin, N., Quessette, F.: An open tool to compute stochastic bounds on steady-state distributions and rewards. In: 11th International Conference on Modeling, Analysis, and Simulation of Computer and Telecommunication Systems, Orlando. IEEE Computer Society (2003)
9. Fourneau, J.M., Le Coz, M., Quessette, F.: Algorithms for an irreducible and lumpable strong stochastic bound. Linear Algebra Appl. **386**, 167–185 (2004)
10. Fourneau, J.M., Pekergin, N.: An algorithmic approach to stochastic bounds. In: Calzarossa, M.C., Tucci, S. (eds.) Performance 2002. LNCS, vol. 2459, pp. 64–88. Springer, Heidelberg (2002). doi:10.1007/3-540-45798-4_4
11. Mitrani, I.: Service center trade-offs between customer impatience and power consumption. Perform. Eval. **68**(11), 1222–1231 (2011)
12. Valmari, A., Franceschinis, G.: Simple $O(m \log n)$ time Markov chain lumping. In: Esparza, J., Majumdar, R. (eds.) TACAS 2010. LNCS, vol. 6015, pp. 38–52. Springer, Heidelberg (2010)

Performance Evaluation

Evaluation of Advanced Routing Strategies with Information-Theoretic Complexity Measures

Michele Amoretti$^{(\boxtimes)}$ and Stefano Cagnoni

University of Parma, Parma, Italy
{michele.amoretti,stefano.cagnoni}@unipr.it

Abstract. Based on hierarchy and recursion (shortly, HR), recursive networking has evolved to become a possible architecture for the future Internet. In this paper, we advance the study of HR-based routing by means of the Gershenson-Fernandez information-theoretic framework, which provides four different complexity measures. Then, we introduce a novel and general approach for computing the information associated to a known or estimated routing table. Finally, we present simulation results regarding networks that are characterized by different topologies and routing strategies. In particular, we discuss some interesting facts we observed while comparing HR-based to traditional routing in terms of complexity measures.

Keywords: Distributed systems · Recursive networking · Complexity measures

1 Introduction

Recursive networking refers to multi-layer virtual networks embedding networks as nodes inside other networks. It is based on *hierarchy*, *i.e.*, the categorization of a set of nodes according to their capability or status, and *recursion*, which is the repeated use of a single functional unit over different scopes of a distributed system. In the last decade, recursive networking has evolved to become a possible architecture for the future Internet [2]. In particular, it is a prominent approach to designing quantum networks [3]. In a recent work [1], we proposed to apply hierarchy and recursion (HR) to build self-aware and self-expressive distributed systems. In particular, we presented HR-based network exploration and routing algorithms.

In this paper, we continue the characterization of HR-based routing by means of a simple albeit powerful and general information-theoretic framework providing complexity measures, recently proposed by Gershenson and Fernandez [4]. Firstly, we introduce a novel and general (*i.e.*, not HR-specific) approach for computing the information associated to a known or estimated routing table. Then we present simulation results regarding networks that are characterized by different topologies and routing strategies. In particular, we discuss some interesting facts we observed, while comparing HR-based to traditional routing in terms of complexity measures.

© The Author(s) 2016
T. Czachórski et al. (Eds.): ISCIS 2016, CCIS 659, pp. 145–153, 2016.
DOI: 10.1007/978-3-319-47217-1_16

The paper is organized as follows. In Sect. 2, we summarize the basic concepts of Gershenson and Fernandez's information-theoretic framework [4]. In Sect. 3, we illustrate our approach for computing the information associated to a routing table. In Sect. 4, we recall the working principles of HR-based routing. In Sect. 5, we present simulation results. Finally, in Sect. 6, we outline future research directions.

2   Complexity and Information

It is difficult to provide an exhaustive list of the ways of defining and measuring system complexity that have been proposed by the research community. Among others, the Gershenson-Fernandez information-theoretic framework provides abstract and concise measures of emergence, self-organization, complexity and homeostasis [4]. According to their framework, emergence is the opposite of self-organization, while complexity represents their balance. Homeostasis can be seen as a measure of the stability of the system.

In detail, a system can be described by a string X, composed by a sequence of variables with values $x \in \{1, .., n\}$ which follow a probability distribution $P(x)$. The information associated to that system is the normalized entropy

$$I = -\frac{\sum_x P(x) \log P(x)}{I_{max}} \tag{1}$$

where $I \in [0, 1]$ and $I_{max} = -\log(1/n)$, since the maximum information value is achieved when all values $1, .., n$ have the same probability.

Considering the dynamics of the system as a process, emergence can be defined as the novel information generated by that process:

$$E = \frac{I}{I_{init}} \tag{2}$$

where I and I_{init} are the current and initial information associated to the system, respectively. The initial information can be referred to the initial state or condition of the system. If the initial state is random, then $I_{init} = 1$.

Self-organization is seen as the opposite of emergence, since high organization (order) is characterized by low information. Vice versa, low organization is characterized by high information. Thus

$$S = I_{init} - I \tag{3}$$

Thus, self-organization occurs ($S > 0$) if the dynamics of the system reduce information.

Since E represents how much variety there is in a system, and S represents how much order, complexity is defined as their product:

$$C = a \cdot E \cdot S \tag{4}$$

where a is a normalization factor, due to the fact that E may be $> 1/S$.

Last but not least, homeostasis is defined as

$$H = 1 - d \tag{5}$$

where d is the normalized *Hamming distance* between the current and initial state of the system, measuring how much change has taken place. Being defined as its complement, homeostasis is a measure of the stability of the system. A high H implies that there is no change, that is, information is maintained.

This framework has been used to study different kinds of complex systems, ranging from self-organizing traffic lights [5] to adaptive peer-to-peer systems [6].

3   Information Associated to a Routing Table

Traditionally, routing strategies are compared in terms of effectiveness, efficiency and scalability [7,8]. To this purpose, selected independent variables should explain performance under a wide range of scenarios [9]. In particular, estimating routing tables is an important and challenging task, as details of how a route is chosen are diverse, and generally not publicly disclosed. An interesting strategy has been recently proposed by Rotenberg *et al.* [10].

In this context, we propose a novel and general approach for characterizing the whole network, namely, by averaging the emergence, self-organization, complexity and homeostasis values of its routers.

From now on, for simplicity, we assume that every node of the network is provided with a routing table, allowing to forward packets to neighbor nodes (routes), according to their destinations. A routing table can be modeled as a set of (destination, route) pairs.

Consider a node with k neighbors. Then, its routing table takes into account k possible routes. In terms of the framework illustrated in Sect. 2, this means that $x \in \{1, .., k\}$. By inspecting the routing table, it is possible to determine the relative frequency of each route. Thus, we define

$$P(x) = \frac{n_x}{n} \tag{6}$$

where n is the size of the routing table and n_x is the number of destinations whose route is x.

When a new node joins the network, its routing table is empty and every route has the same probability. Thus, $I_{init} = 1$. As a consequence, Eqs. 2–5 become:

- $E = I/I_{init} = I$
- $S = I_{init} - I = 1 - I$
- $C = aES = 4I(1 - I)$
- $H = 1 - d$

where $a = 4$ comes from: $\max\{ES\} = 0.5(1 - 0.5) = 1/4$; d is the normalized Hamming distance between the initial and current configurations of the routing table. In general, the Hamming distance between any two consecutive configurations of the routing table is computed per-node according to the following equations:

$$\forall \text{ neighbor } i :$$
$$D_{i+1} = D_i + f(r)$$
$$f(i) = \begin{cases} 1 & \text{if } r \text{ is route in new routing table only} \\ 1 & \text{if } r \text{ is route in old routing table only} \\ 1 & \text{if } r \text{ is route in both routing tables, but } n_{old} \neq n_{new} \\ 0 & \text{else} \end{cases}$$

where n is the number of destinations associated to the selected route. Once normalized, D_i becomes d_i.

4   HR-Based Routing

We recall and explain HR-based routing by means of an example. Let us consider the network shown in Fig. 1. The routing table at node 4.2 contains information on how to reach any other node in the network. The table has more precise information about nearby destinations (node 4.4 and node 4.7), and vague information about more remote destinations (NET9).

Suppose that node 4.2 has to send a message to node 9.6. If routing tables were filled only with local information (*i.e.*, node 4.2's direct neighbors), routing would be quite inefficient. Instead, hierarchy and recursion make it possible to find the route more quickly. Node 4.2 knows that NET9 is reachable through NET6, whose node 6.1 is directly reachable. Thus, node 4.2 sends the message to node 6.1. The complete HR-based routing algorithm is described by the flowchart in Fig. 2.

HR-based routing is suitable for both intra-domain and inter-domain scenarios. Compared to the two main classes of intra-domain routing, namely Link-State and Distance-Vector [7], HR-based routing has the following advantages:

1. nodes are not required to know the whole network topology (unlike Link-State routing);
2. nodes build collective awareness by exchanging recursive and hierarchical information not only with direct neighbors, but also with neighbors of neighbors, etc. (unlike Distance-Vector routing).

For further details about HR-based versus Link-State and Distance-Vector, the reader may refer to our previous work [1]. Thanks to collective awareness, messages can be routed within the same subnetwork or from one subnetwork to another; doing so they enable, for example, the Unified Architecture for inter-domain routing proposed in RFC 1322.[1]

---

[1] http://www.rfc-editor.org/rfc/rfc1322.txt.

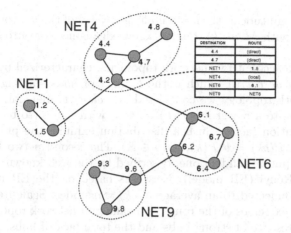

Fig. 1. Hierarchy and recursion: the routing table at node 4.2 contains information on how to reach any other node in the network.

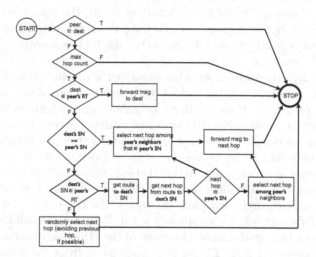

Fig. 2. HR-based routing algorithm. RT stands for routing table; SN for subnetwork.

5   Simulation Results

To evaluate the proposed approach, we used the general-purpose discrete event simulation environment DEUS [11]. The purpose of DEUS is to facilitate the simulation of highly dynamic overlay networks with several hundred thousands nodes, without needing to simulate also lower network layers.

Without loss of generality, we considered the (sub-optimal) scenario in which every node knows which subnetworks can be reached through its direct neighbors. In HR-based routing, no further knowledge — provided by neighbors of neighbors (of neighbors etc.) — is necessary, when the number of subnetworks M is of the same order of magnitude as the mean node degree $\langle k \rangle$ of the network. Instead, for large networks, with $M \gg \langle k \rangle$, further knowledge is necessary to build effective routing tables.

We took into account two network topologies, characterized by different statistics for the *node degree*, which is the number of links starting from a node. The first network topology we considered is *scale-free*, meaning that its PMF decays according to a power law $P(k) = ck^{-\tau}$, with $\tau > 1$ (to be normalizable) and c normalization factor. Such a distribution exhibits the property of scale invariance (*i.e.*, $P(bk) = b^a P(k)$, $\forall a, b \in \mathbb{R}$). The second network topology we considered is a purely-random one, described by the well-known model defined by Erdös and Rényi (ER model). Networks based on the ER model have N vertices, each connected to an average of $\langle k \rangle = \alpha$ nodes. Scale-free and purely-random are the extremes of the range of meaningful network topologies, as they represent the presence of strong hubs and the total lack of hubs, respectively.

We evaluated the HR-based routing strategy in terms of success rate (*i.e.*, fraction of messages arrived to destination) and average route length, using different networks characterized by $N = 1000$ nodes, with $M = 20$ subnetworks. With the BA topology, when $m = 5$ and $m = 20$, the mean node degree is $\langle k \rangle = 10$ and $\langle k \rangle = 40$, respectively. To have the same $\langle k \rangle$ values for the ER topology, we set $\alpha = 10$ and $\alpha = 40$. Reported results are average values coming from 25 simulation runs.

As a basis for comparison, we also simulated a routing strategy where the nodes do not populate routing tables with information about subnetworks. Instead, they only keep trace of direct neighbors and neighbors of neighbors. Such a strategy (denoted as No-HR) has some common properties with Distance-Vector routing, although it does not manipulate vectors of distances to other nodes in the network. Mean values and standard deviations of success rate r_s and average route length n_h, reported in Table 1, show that the HR-based strategy outperforms the other one, provided that the average node degree $\langle k \rangle$ is suitably high. Interestingly, with low $\langle k \rangle$ values, the HR-based routing strategy has worse performance when the topology is ER. However, a small increase of $\langle k \rangle$ corresponds to a high performance increase of the HR-based routing strategy.

Then, we computed E, S, C and H at each node, from the initial configuration corresponding to $I_{init} = 1$, to the steady-state configuration corresponding to the filled routing table. We averaged the resulting values, considering the whole network. Their evolution is illustrated in Fig. 3.

Four main facts can be observed:

1. As m and α grow, E tends to 1, S tends to 0.
2. When m and α are low, HR-based and NoHR routing exhibit very different H values.
3. When m and α are high, the values of H in HR-based and NoHR routing are more similar.
4. Even if the mean node degree $\langle k \rangle$ is the same, BA and ER topologies result in very different E, S and C values.

Table 1. HR vs NoHR: success rate r_s and average route length n_h

Strategy	Topology	S	μ_{r_s}	σ_{r_s}	μ_{n_h}	σ_{n_h}
HR	BA, $m = 5$	20	0.88	2E-2	17.6	2.06
NoHR	BA, $m = 5$	20	0.74	2.9E-1	19.7	9.8
HR	BA, $m = 20$	20	0.99	9.3E-4	3.8	8E-2
NoHR	BA, $m = 20$	20	0.99	9E-3	9.85	1.2
HR	ER, $\alpha = 10$	20	0.64	3E-2	43.7	2.72
NoHR	ER, $\alpha = 10$	20	0.55	3.3E-1	21.64	17.91
HR	ER, $\alpha = 40$	20	0.99	1E-3	4.0	1.2E-1
NoHR	ER, $\alpha = 40$	20	0.93	1.9E-1	15.33	4.75

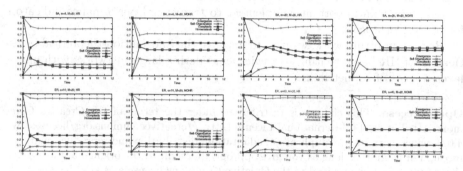

Fig. 3. Complexity measures of HR-based and NoHR routing with different topologies.

The reason for the first fact is that a higher number of connections, due to higher m and α, makes the routing table more varied in terms of available routes. The probability distribution $P(x)$ has fewer spikes, thus I is higher. As a consequence, E increases and S decreases. The second fact can be stated more precisely by means of the following inequality: $H_{HR} \ll H_{NoHR}$, when m and α are small. Our interpretation is that a reduced number of connections enhances the differences between routing tables, in HR-based and NoHR routing, *i.e.*, with respect to the initial state, the final state of the routing table is much more different in HR-based routing rather than NoHR routing. The impact on performance is evident: HR routing table are better than NoHR ones, thus producing a higher success rate. It is not possible, however, to generalize associating higher H values to higher performance. Conversely, a higher number of connections reduces the differences between routing tables, explaining the third fact. The fourth fact is further detailed by the following inequalities: $E_{BA} < E_{ER}$, $S_{BA} > S_{ER}$ and $C_{BA} \gg C_{ER}$, when m and α are such that the mean node degree $\langle k \rangle$ is the same in the BA and ER topologies. It is difficult to explain the relationship between C and performance, in general. It makes more sense to consider E and S separately. Regarding E, our interpretation is that scale-free properties (characterizing the BA topology) make some routes intrinsically more probable than others. Indeed, only a few nodes have a high number of connections (such nodes are denoted as *hubs*). Thus, with respect to the ER topology, in scale-free networks the probability distribution $P(x)$ has more spikes, making I smaller. Consequently, E is lower and S is higher. Indeed, the presence of hubs makes routing more robust (S is higher), thus improving performance.

6   Conclusion

In this paper we have illustrated a novel approach to quantifying the information associated to a known or estimated routing table, allowing to characterize the whole network by averaging the emergence, self-organization, complexity and homeostasis values of its nodes. Our simulation study shows that these measures may represent an important complement to traditional performance indicators for routing protocols.

Regarding future work, we plan to improve the information-theoretical investigation of HR-based routing strategies, considering larger networks with multi-layered trees of subnetworks.

References

1. Amoretti, M., Cagnoni, S.: Toward collective self-awareness and self-expression in distributed systems. IEEE Comput. **48**(7), 29–36 (2015)
2. Touch, J., Baldine, I., Dutta, R., Ford, B., Finn, G., Jordan, S., Massey, D., Matta, A., Papadopoulos, C., Reiher, P., Rouskas, G.: A dynamic recursive unified internet design (DRUID). Comput. Netw. **55**(4), 919–935 (2011)
3. Van Meter, R.: Quantum networking and internetworking. IEEE Netw. **26**, 59 (2012)
4. Gershenson, C., Fernandez, N.: Complexity and information: measuring emergence. Self Organ. Homeost. Multiple Scales Complex. **18**(2), 29–44 (2012)
5. Zubillaga, D., Cruz, G., Aguilar, L.D., Zapotecatl, J., Fernandez, N., Aguilar, J., Rosenblueth, D.A., Gershenson, C.: Measuring the complexity of self-organizing traffic lights. Entropy **16**(5), 2384–2407 (2014)
6. Amoretti, M., Gershenson, C.: Measuring the complexity of adaptive peer-to-peer systems. Peer Peer Network. Appl. **9**, 1031 (2016)

7. Kurose, J.F., Ross, K.W.: Computer Networking: A Top-Down Approach. Addison-Wesley, Boston (2012)
8. Akkaya, K., Younis, M.: A survey on routing protocols for wireless sensor networks. Ad Hoc Netw. **3**(3), 325–349 (2005)
9. Stojmenovic, I.: Simulations in wireless sensor and ad hoc networks: matching and advancing models, metrics, and solutions. IEEE Commun. Mag. **46**(12), 102–107 (2008)
10. Rotenberg, É., Crespelle, C., Latapy, M.: Measuring routing tables in the internet. In: 6th IEEE International Workshop on Network Science for Communication Networks, Toronto, Canada (2014)
11. Amoretti, M., Picone, M., Zanichelli, F., Ferrari, G.: Simulating mobile and distributed systems with DEUS and ns-3. In: HPCS, Helsinki, Finland (2013)

Performance of Selection Hyper-heuristics on the Extended HyFlex Domains

Alhanof Almutairi[1]([✉]), Ender Özcan[1], Ahmed Kheiri[2],
and Warren G. Jackson[1]

[1] ASAP Research Group, School of Computer Science, University of Nottingham,
Wollaton Road, Nottingham NG8 1BB, UK
{psxaka,ender.ozcan,psxwgj}@nottingham.ac.uk
[2] Operational Research Group, School of Mathematics, Cardiff University,
Senghennydd Road, Cardiff CF24 4AG, UK
KheiriA@cardiff.ac.uk

Abstract. Selection hyper-heuristics perform search over the space of heuristics by mixing and controlling a predefined set of low level heuristics for solving computationally hard combinatorial optimisation problems. Being reusable methods, they are expected to be applicable to multiple problem domains, hence performing well in cross-domain search. HyFlex is a general purpose heuristic search API which separates the high level search control from the domain details enabling rapid development and performance comparison of heuristic search methods, particularly hyper-heuristics. In this study, the performance of six previously proposed selection hyper-heuristics are evaluated on three recently introduced extended HyFlex problem domains, namely 0–1 Knapsack, Quadratic Assignment and Max-Cut. The empirical results indicate the strong generalising capability of two adaptive selection hyper-heuristics which perform well across the 'unseen' problems in addition to the six standard HyFlex problem domains.

Keywords: Metaheuristic · Parameter control · Adaptation · Move acceptance · Optimisation

1 Introduction

Many combinatorial optimisation problems are computationally difficult to solve and require methods that use sufficient knowledge of the problem domain. Such methods cannot however be reused for solving problems from other domains. On the other hand, researchers have been working on designing more general solution methods that aim to work well across different problem domains. Hyper-heuristics have emerged as such methodologies and can be broadly categorised into two categories; *generation* hyper-heuristics to generate heuristics from existing components, and *selection* hyper-heuristics to select the most appropriate heuristic from a set of low level heuristics [3]. This study focuses on selection hyper-heuristics.

© The Author(s) 2016
T. Czachórski et al. (Eds.): ISCIS 2016, CCIS 659, pp. 154–162, 2016.
DOI: 10.1007/978-3-319-47217-1_17

A selection hyper-heuristic framework operates on a single solution and iteratively *selects* a heuristic from a set of low level heuristics and applies it to the candidate solution. Then a *move acceptance* method decides whether to accept or reject the newly generated solution. This process is iteratively repeated until a termination criterion is satisfied. In [5], a range of simple selection methods are introduced, including *Simple Random* (SR) that randomly selects a heuristic at each step, and *Random Descent* which works similarly to SR, but the selected low level heuristic is applied repeatedly until no additional improvement in the solution is observed. Most of the simple non-stochastic basic move acceptance methods are tested in [5]; including *All Moves* (AM), which accepts all moves, *Only Improving* (OI), which accepts only improving moves and *Improving or Equal* (IE), which accepts all non-worsening moves. *Late acceptance* [4] accepts an incumbent solution if its quality is better than a solution that was obtained a specific number of steps earlier. More on selection hyper-heuristics can be found in [3].

HyFlex [14] (**Hyper**-heuristics **Flex**ible framework) is a cross-domain heuristic search API and HyFlex v1.0 is a software framework written in Java, providing an easy-to-use interface for the development of selection hyper-heuristic search algorithms along with the implementation of several problem domains, each of which encapsulates problem-specific components, such as solution representation and low level heuristics. We will refer to HyFlex v1.0 as HyFlex from this point onward. HyFlex was initially developed to support the first Cross-domain Heuristic Search Challenge (CHeSC) in 2011[1]. Initially, there were six minimisation problem domains implemented within HyFlex [14]. The HyFlex problem domains have been extended to include three more of them, including 0–1 Knapsack Problem (KP), Quadratic Assignment Problem (QAP) and Max-Cut (MAC) [1]. In this study, we only consider the 'unseen' extended HyFlex problem domains to investigate the performance and the generality of some previously proposed well performing selection hyper-heuristics.

2 Selection Hyper-heuristics for the Extended HyFlex Problem Domains

In this section, we provide a description of the selection hyper-heuristic methods which are investigated in this study. These hyper-heuristics use different combinations of heuristic selection and move acceptance methods.

Sequence-based selection hyper-heuristic (SSHH) [10] is a relatively new method which aims to discover the best performing sequences of heuristics for improving upon an initially generated solution. The hidden Markov model (HMM) is employed to learn the optimum sequence lengths of heuristics. The hidden states in HMM are replaced by the low level heuristics and the observations in HMM are replaced by the sequence-based acceptance strategies (AS). A transition probabilities matrix is utilised to determine the movement between the hidden states; and an emission probabilities matrix is employed to determine

---

[1] http://www.asap.cs.nott.ac.uk/external/chesc2011/.

whether a particular sequence of heuristics will be applied to the candidate solution or will be coupled with another LLH. The move acceptance method used in [10] accepts all improving moves and non-improving moves with an adaptive threshold. The SSHH showed excellent performance across CHeSC 2011 problem domains achieving better overall performance than Adap-HH which was the winner of the challenge.

Dominance-based and random descent hyper-heuristic (DRD) [16] is an iterated multi-stage hyper-heuristic that hybridises a dominance-based and random descent heuristic selection strategies, and uses a naïve move acceptance method which accepts improving moves and non-improving moves with a given probability. The dominance-based stage uses a greedy-like method aiming to identify a set of 'active' low level heuristics considering the trade-off between the delta change in the fitness and the number of iterations required to achieve that change. The random descent stage considers only the subset of low level heuristics recommended by the dominance-based stage. If the search stagnates, then the dominance-based stage may kick in again aiming to detect a new subset of active heuristics. The method has proven to perform relatively well in the MAX-SAT and 1D bin-packing problem domains as reported in [16].

Robinhood (*round-robin* neighbour*hood*) hyper-heuristic [11] is an iterated multi-stage hyper-heuristic. Robinhood contains three selection hyper-heuristics. They all share the same heuristic selection method but differ in the move acceptance. The Robinhood heuristic selection allocates equal time for each low level heuristic and applies them one at a time to the incumbent solution in a cyclic manner during that time. The three move acceptance criteria employed by Robinhood are only improving, improving or equal, and an adaptive move acceptance method. The latter method accepts all improving moves and non-improving moves are accepted with a probability that changes adaptively throughout the search process. This selection hyper-heuristic outperformed eight 'standard' hyper-heuristics across a set of instances from HyFlex problem domains. A detailed description of the Robinhood hyper-heuristic can be found in [11].

Modified choice function (MCF) [6] uses an improved version of the traditional choice function (CF) heuristic selection method used in [5] and has a better average performance than CF when compared across the CHeSC 2011 competition problems. The basic idea of a choice function hyper-heuristic is to choose the best low level heuristic at each iteration. Hence, move acceptance is not needed and all moves are accepted. In the traditional CF method, each low level heuristic is assigned a score based on three factors; the recent effectiveness of the given heuristic (f_1), the recent effectiveness of consecutive pairs of heuristics (f_2), and the amount of time since the given heuristic was used (f_3) where each factor within CF is associated with a weight; α, β, and δ respectively [5]. It was also stated in the CF study that the hyper-heuristic was insensitive to the parameter settings for solving Sales Summit Scheduling problems and are consequently fixed throughout the search. MCF extends upon CF by controlling the weights of each factor for improving its cross-domain performance [6]. In MCF, the weights for f_1 and f_2 are equal as defined by the parameter ϕ_t, and the weight for f_3 is set to $1 - \phi_t$. ϕ_t is controlled using a simple mechanism. If an improving move is made, then $\phi_t = 0.99$. If a non-improving move is made, then $\phi_t = max\{\phi_{t-1} - 0.01, 0.01\}$.

Fuzzy late acceptance-based hyper-heuristic (F-LAHH) [8] was implemented for solving MAX-SAT problems and showed promising results. F-LAHH utilises a fitness proportionate selection mechanism (RUA1-F1FPS) [7] for the heuristic selection method and uses late acceptance, whose list length is adaptively controlled using a fuzzy control system, for its move acceptance method. In RUA1-F1FPS, the low level heuristics are assigned scores which are updated based on acceptance of the candidate solution as defined by the RUA1 scheme. A heuristic is chosen using a fitness proportionate (roulette wheel) selection mechanism utilising Formula 1 (F1) ranking scores (F1FPS). Each low level heuristic is ranked based on their current scores using F1 ranking and are assigned probabilities to be selected proportional to their F1 rank. The fuzzy control system, as defined in [8], adapts the list length of a late acceptance move acceptance method at the start of each phase each to promote intensification or diversification within the subsequent phase of the search based on the amount of improvement over the current phase. The F1FPS scoring mechanism used in this study is the RUA1 method as used in [7,8]. The parameters of the fuzzy system are the same as those used in [8] with the universe of discourse of the list length fuzzy sets $U = [10000, 30000]$, the initial list length of late acceptance $L_0 = 10000$, and the number of phases equal to 50.

Simple Random-Great Deluge (SR-GD) is a single-parameter selection hyper-heuristic method. At each step, a random heuristic will be selected and applied to the current solution. Great deluge move acceptance method [9] accepts improving solutions by default. A non-improving solution is only accepted if its quality is better than a threshold level at each iteration. Initially, the threshold level is set to the cost of the initially constructed solution. The threshold level is then updated at each iteration with a linear rate given by the following formula:

$$T_t = c + \Delta C \times \left(1 - \frac{t}{N}\right) \tag{1}$$

where T_t is the value of the threshold level at time t, N is the time limit, ΔC is the expected range for the maximum change in the cost, and c is the final cost.

3   Empirical Results

The methods presented in Sect. 2 are applied to 10 instances from each of the recently introduced HyFlex problem domains. The experiments are conducted on an i7-3820 CPU at 3.60 GHz with a memory of 16.00 GB. Each run is repeated 31 times with a termination criteria of 415 s corresponding to 600 nominal seconds of the CHeSC 2011 challenge test machine[2]. The following performance indicators are used for ranking hyper-heuristics across all three domains:

---

[2] http://www.asap.cs.nott.ac.uk/external/chesc2011/benchmarking.html.

- **rank:** rank of a hyper-heuristic with respect to μ_{norm}.
- μ_{rank}: each algorithm is ranked based on the median objective values that they produce over 31 runs for each instance. The top algorithm is assigned to rank 1, while the worst algorithm's rank equals to the number of algorithms being considered in ranking. In case of a tie, the ranks are shared by taking the average. The ranks are then accumulated and averaged over all instances producing μ_{rank}.
- μ_{norm}: the objective function values are normalised to values in the range [0,1] based on the following formula:

$$norm(o, i) = \frac{o(i) - o_{best}(i)}{o_{worst}(i) - o_{best}(i)} \tag{2}$$

where $o(i)$ is the objective function value on instance i, $o_{best(i)}$ is the best objective function value obtained by all methods on instance i, and $o_{worst(i)}$ is the worst objective function value obtained by all methods on instance i. μ_{norm} is the average normalised objective function value.

- **best:** is the number of instances for which the hyper-heuristic achieves the best median objective function value.
- **worst:** the number of instances for which the hyper-heuristic delivers the worst median objective function value.

As a performance indicator, μ_{rank} focusses on median values and does not consider how far those values are from each other for the algorithms in consideration, while μ_{norm} considers the mean performance of algorithms by taking into account the relative performance of all algorithms over all runs across each problem instance.

Table 1 summarises the results. On KP, SSHH delivers the best median values for 8 instances including 4 ties. Robinhood achieves the best median results in 5 instances including a tie. SR-GD, F-LAHH and DRD show comparable performance. On the QAP problem domain, SR-GD performs the best in 6 instances and F-LAHH shows promising results in this particular problem domain. This gives an indication that simple selection methods are potentially the best for solving QAP problems. SSHH ranked as the third best based on the average rank on QAP problem. On MAC, SSHH clearly outperforms all other methods, followed by SR-GD and then Robinhood. The remaining hyper-heuristics have relatively poor performance, with MCF being the worst of the 6 hyper-heuristics. Overall, SSHH turns out to be the best with $\mu_{norm} = 0.16$ and $\mu_{rank} = 2.28$. SR-GD also shows promising performance, scoring the second best. MCF consistently delivers weak performance in all the instances of the three problem domains. Table 1 also provides the pairwise average performance comparison of SSHH versus (DRD, Robinhood, MCF, F-LAHH and SR-GD) based on the Mann-Whitney-Wilcoxon statistical test. SSHH performs significantly better than any hyper-heuristic on all MAC instances, except Robinhood which performs better than SSHH on four out of ten instances. On the majority of the KP instances, SSHH is the best performing hyper-heuristic. SSHH performs poorly on QAP

Table 1. The performance comparison of SSHH, DRD, Robinhood, MCF, F-LAHH and SR-GD over 31 runs for each instance. The best median values per each instance are highlighted in bold. Based on the Mann-Whitney-Wilcoxon test, for each pair of algorithms; SSHH versus X; SSHH > (<) X indicates that SSHH (X) is better than X (SSHH) and this performance variance is statistically significant with a confidence level of 95%, and SSHH ≥ (≤) X indicates that there is no statistical significant between SSHH and X, but SSHH (X) is better than X (SSHH) on average.

Domain Instance	SSHH med. rank	SSHH min. vs	DRD med. rank	DRD min. vs	Robinhood med. rank	Robinhood min. vs	MCF med. rank	MCF min. vs	F-LAHH med. rank	F-LAHH min. vs	SR-GD med. rank	SR-GD min.
KP												
Inst1	**-104046** 1	-104046 >	-104025 4.5	-104044 >	-104034 3	-104046 >	-103998 6	-104046 >	-104037 2	-104046 >	-104025 4.5	-104046
Inst2	**-1247642** 1	-1261320 ≥	-1208666 6	-1208666 ≥	-1241628 5	-1253664 >	-1226625 3	-1244413 >	-1212253 2	-1220422 >	-1212829 4	-1221623
Inst3	**-241934** 1	-242963 >	-232525 6	-233066 >	-236420 5	-238447 >	-239323 2	-240023 >	-238397 4	-239848 >	-238664 3	-239192
Inst4	**-431350** 1	-431362 >	-431333 3	-431349 >	-431320 5	-431338 >	-431325 2	-431341 >	-431314 6	-431331 >	-431316 5	-431329
Inst5	**-396167** 3	-396167 ≤	-396167 3	-396167 ≤	-396167 3	-396167 ≤	-396127 6	-396167 ≤	-396167 3	-396167 ≤	-396167 3	-396167
Inst6	-4251693 4	-4268665 >	-4248962 5.5	-4248962 >	**-4262735** 1	-4312111 >	-4248962 5.5	-4321660 ≤	-4251867 3	-4268839 >	-4253175 2	-4273295
Inst7	-929052 2	-943136 >	-924303 5	-924357 >	-924346 4	-933892 >	-923904 6	-939879 ≤	-924937 3	-941397 >	**-935411** 1	-940485
Inst8	**-1577175** 2.5	-1577175 >	-1577166 5	-1577175 >	**-1577175** 2.5	-1577175 >	-1572999 6	-1572999 ≥	**-1577175** 2.5	-1577175 ≤	**-1577175** 2.5	-1577175
Inst9	**-1530477** 1.5	-1530511 >	-1530465 3.5	-1530485 >	**-1530477** 1.5	-1530494 >	-1530465 3.5	-1530498 >	-1530453 5.5	-1530484 >	-1530453 5.5	-1530463
Inst10	**-1467357** 2	-1467362 >	-1467357 2	-1467357 >	-1467357 1.5	-1467362 >	-1457070 6	-1457353 >	-1467353 4.5	-1467361 >	-1467353 4.5	-1467362
rank average	1.90		4.25		2.80		4.70		3.85		3.50	
norm average	0.10		0.27		0.17		0.30		0.25		0.17	
QAP												
Inst1	152572 4	152224 ≤	**152000** 5	152000 ≤	152686 5	152334 >	153398 6	152700 >	152372 3	152122 >	152258 2	152068
Inst2	154492 3	154130 ≥	155000 5	155000 ≥	154616 4	154136 >	155300 6	154706 >	154178 2	153960 >	**154172** 1	154016
Inst3	148374 3	147930 ≥	148604 4	147916 ≤	148462 4	148088 >	149584 6	148604 >	148140 2	148026 >	**148056** 1	147900
Inst4	150366 4	149782 ≤	150336 3	149724 ≤	150380 4	150002 >	151016 6	150164 >	149978 2	149730 >	**149892** 1	149688
Inst5	21419490 4	21325030 >	21400000 3	21300000 >	21383596 2	21325716 >	21598704 6	21414834 >	21495226 5	21351226 >	**21361794** 1	21207680
Inst6	1190346287 4	1186663179 ≥	1190000000 3	1190000000 ≥	1199401744 5	1192546366 >	1204968089 6	1204968089 >	1188454126 2	1186678730 >	**1188111647** 1	1186811188
Inst7	504406437 4	500015697 ≥	504000000 3	502000000 ≥	508225133 5	504102563 >	506396735 6	506396735 >	**501945504** 1	500096792 >	502027073 2	500227073
Inst8	44892452 3	44855568 >	44900000 4	44800000 >	44933092 6	44875514 ≥	44903670 5	44869704 ≥	**44859724** 1	44841194 >	44863858 2	44842660
Inst9	8179752 3	8151040 >	8200846 4	8165384 >	8202996 5	8177206 >	8254190 6	8213094 >	**8162896** 1	8157314 >	8163776 2	8150316
Inst10	273622 3	273216 >	274000 5	273000 >	273908 4	273590 >	274404 6	273566 >	273460 2	273264 ≤	**273362** 1	273216
rank average	3.50		3.60		4.50		5.90		2.10		1.40	
norm average	0.24		0.29		0.32		0.58		0.16		0.12	
MAC												
Inst1	**-41101646** 1	-41517765 >	-39393891 1	-40202568 >	-40471041 1	-40863976 >	-40157605 5	-40967725 >	-40419083 4	-41268393 >	-40756746 2	-41377263
Inst2	**-273938900** 1	-277548425 >	-266329920 5	-268635140 >	-269502099 2	-271045451 >	-256423018 5	-256442443 >	-266773056 4.5	-274334343 >	-267482996 3	-269292120
Inst3	**-3056** 1	-3062 >	-3014 6	-3030 >	-3043 4.5	-3051 >	-3046 3	-3056 >	-3043 4.5	-3053 >	-3053 3	-3057
Inst4	**-3040** 1	-3050 >	-2991 6	-3012 >	-3027 4.5	-3032 >	-3027 3	-3033 >	-3027 4.5	-3035 >	-3035 3	-3047
Inst5	**-3041** 1	-3051 >	-3000 6	-3016 >	-3028 4.5	-3034 >	-3029 3	-3042 >	-3028 4.5	-3042 >	-3038 3	-3045
Inst6	**-13243** 2	-13300 >	-13047 6	-13106 >	-13204 4.5	-13246 >	-13176 5	-13241 >	-13186 4	-13247 >	-13216 3	-13284
Inst7	-1352 2	-1358 >	-1246 6	-1278 >	**-1362** 1	-1368 ≤	-1316 5	-1330 >	-1322 4	-1342 >	-1334 3	-1346
Inst8	-10074 2	-10125 >	-9819 6	-9872 >	**-10152** 1	-10190 ≤	-9964 5	-9996 >	-10004 4	-10101 >	-10046 3	-10078
Inst9	-454 2.5	-458 >	-416 6	-430 ≥	-454 2.5	-456 >	-444 5	-454 >	-440 4	-450 >	**-456** 1	-456
Inst10	-2912 2	-2960 >	-2676 6	-2704 >	**-2942** 1	-2952 >	-2810 5	-2842 >	-2848 4	-2906 >	-2884 3	-2926
rank average	1.45		5.90		2.65		4.50		4.20		2.30	
norm average	0.14		0.74		0.21		0.40		0.34		0.22	
H_{rank}	2.28		4.58		3.32		5.03		3.38		2.40	
H_{norm}	0.16		0.43		0.23		0.43		0.25		0.17	

Table 2. The performance comparison of SSHH, Adap-HH, FS-ILS, NR-FS-ILS, EPH, SR-AM and SR-IE

KP Problem Domain						QAP Problem Domain				
rank	method	μ_{rank}	μ_{norm}	best	worst	rank	method	μ_{rank}	μ_{norm}	best	worst
1	Adap-HH	1.95	0.027	8	0	1	NR-FS-ILS	1.95	0.100	5	0
2	EPH	2.35	0.053	4	0	2	Adap-HH	2.50	0.103	2	0
3	SSHH	2.45	0.059	5	0	3	FS-ILS	2.85	0.103	3	0
4	SR-AM	4.40	0.148	2	0	4	EPH	3.80	0.133	0	0
5	SR-IE	5.55	0.328	0	4	5	SR-AM	4.10	0.146	1	0
6	NR-FS-ILS	5.60	0.361	1	6	6	SSHH	5.80	0.189	0	0
7	FS-ILS	5.70	0.395	1	2	7	SR-IE	7.00	0.634	0	10

MAC Problem Domain						Overall				
rank	method	μ_{rank}	μ_{norm}	best	worst	rank	method	μ_{rank}	μ_{norm}	best	worst
1	SSHH	1.35	0.092	9	0	1	SSHH	3.20	0.113	14	0
2	SR-AM	2.45	0.252	1	0	2	Adap-HH	2.53	0.135	10	0
3	Adap-HH	3.15	0.275	0	0	3	SR-AM	3.65	0.182	4	0
4	NR-FS-ILS	4.00	0.374	0	0	4	EPH	3.92	0.235	4	1
5	FS-ILS	4.85	0.392	1	2	5	NR-FS-ILS	3.85	0.278	6	6
6	EPH	5.60	0.519	0	1	6	FS-ILS	4.47	0.297	5	4
7	SR-IE	6.60	0.732	0	7	7	SR-IE	6.38	0.565	0	21

when compared to F-LAHH and SR-GD and both hyper-heuristics produce significantly better results than SSHH on almost all instances. SSHH performs statistically significantly better than the remaining hyper-heuristics on QAP.

The performance of the best hyper-heuristic from Table 1, SSHH is compared to the methods whose performances are reported in [1], including Adap-HH, which is the winner of the CHeSC 2011 competition [13], an Evolutionary Programming Hyper-heuristic (EPH) [12], Fair-Share Iterated Local Search with (FS-ILS) and without restart (NS-FS-ILS), Simple Random-All Moves (SR-AM) (denoted as AA-HH previously) and Simple Random-Improving or Equal (SR-IE) (denoted as ANW-HH previously). Table 2 summarises the results based on μ_{rank}, μ_{norm}, best and worst counts. Adap-HH performs better than SSHH in KP and QAP while SSHH performs the best on MAC. Overall, SSHH is the best method based on μ_{norm} with a value of 0.113, however Adap-HH is the top ranking algorithm based on μ_{rank} with a value of 2.53 and SSHH is the second best with a value of 3.20.

4   Conclusion

A hyper-heuristic is a search methodology, designed with the aim of reducing the human effort in developing a solution method for multiple computationally difficult optimisation problems via automating the mixing and generation of heuristics. The goal of this study was to assess the level of generality of a set of selection hyper-heuristics across three recently introduced HyFlex problem domains. The empirical results show that both Adap-HH and SSHH perform better than the previously proposed algorithms across the problem domains included in the HyFlex extension set. Both adaptive algorithms embed different online learning mechanisms and indeed generalise well on the 'unseen' problems. It has also been observed that the choice of heuristic selection and move acceptance combination could lead to major performance differences across a diverse set of problem domains. This particular observation is aligned with previous findings in [2,15].

References

1. Adriaensen, S., Ochoa, G., Nowé, A.: A benchmark set extension and comparative study for the HyFlex framework. In: Proceedings of IEEE Congress on Evolutionary Computation, pp. 784–791 (2015)
2. Bilgin, B., Özcan, E., Korkmaz, E.E.: An experimental study on hyper-heuristics and exam timetabling. In: Burke, E.K., Rudová, H. (eds.) PATAT 2006. LNCS, vol. 3867, pp. 394–412. Springer, Heidelberg (2007). doi:10.1007/978-3-540-77345-0_25
3. Burke, E.K., Gendreau, M., Hyde, M., Kendall, G., Ochoa, G., Özcan, E., Qu, R.: Hyper-heuristics: a survey of the state of the art. J. Oper. Res. Soc. **64**(12), 1695–1724 (2013)
4. Burke, E.K., Bykov, Y.: A late acceptance strategy in hill-climbing for exam timetabling problems. In: Proceedings of the 7th International Conference on the Practice and Theory of Automated Timetabling (PATAT 2008) (2008)
5. Cowling, P.I., Kendall, G., Soubeiga, E.: A hyperheuristic approach to scheduling a sales summit. In: Burke, E., Erben, W. (eds.) PATAT 2000. LNCS, vol. 2079, p. 176. Springer, Heidelberg (2001)
6. Drake, J.H., Özcan, E., Burke, E.K.: An improved choice function heuristic selection for cross domain heuristic search. In: Coello, C.A.C., Cutello, V., Deb, K., Forrest, S., Nicosia, G., Pavone, M. (eds.) PPSN 2012, Part II. LNCS, vol. 7492, pp. 307–316. Springer, Heidelberg (2012)
7. Jackson, W.G., Özcan, E., Drake, J.H.: Late acceptance-based selection hyper-heuristics for cross-domain heuristic search. In: 13th UK Workshop on Computational Intelligence, pp. 228–235 (2013)
8. Jackson, W., Özcan, E., John, R.I.: Fuzzy adaptive parameter control of a late acceptance hyper-heuristic. In: 14th UK Workshop on Computational Intelligence (UKCI), pp. 1–8 (2014)

9. Kendall, G., Mohamad, M.: Channel assignment optimisation using a hyper-heuristic. In: Proceedings of the IEEE Conference on Cybernetic and Intelligent Systems, pp. 790–795 (2004)
10. Kheiri, A., Keedwell, E.: A sequence-based selection hyper-heuristic utilising a hidden Markov model. In: Proceedings of the 2015 on Genetic and Evolutionary Computation Conference, GECCO 2015, pp. 417–424. ACM, New York (2015)
11. Kheiri, A., Özcan, E.: A hyper-heuristic with a round Robin neighbourhood selection. In: Middendorf, M., Blum, C. (eds.) EvoCOP 2013. LNCS, vol. 7832, pp. 1–12. Springer, Heidelberg (2013)
12. Meignan, D.: An evolutionary programming hyper-heuristic with co-evolution for CHeSC11. In: The 53rd Annual Conference of the UK Operational Research Society (OR53) (2011)
13. Misir, M., Verbeeck, K., De Causmaecker, P., Vanden Berghe, G.: A new hyper-heuristic implementation in HyFlex: a study on generality. In: Fowler, J., Kendall, G., McCollum, B. (eds.) Proceedings of the 5th Multidisciplinary International Scheduling Conference: Theory and Application (MISTA2011), pp. 374–393 (2011)
14. Ochoa, G., Hyde, M., Curtois, T., Vazquez-Rodriguez, J.A., Walker, J., Gendreau, M., Kendall, G., McCollum, B., Parkes, A.J., Petrovic, S., Burke, E.K.: HyFlex: a benchmark framework for cross-domain heuristic search. In: Hao, J.-K., Middendorf, M. (eds.) EvoCOP 2012. LNCS, vol. 7245, pp. 136–147. Springer, Heidelberg (2012)
15. Özcan, E., Bilgin, B., Korkmaz, E.E.: A comprehensive analysis of hyper-heuristics. Intell. Data Anal. 12(1), 3–23 (2008)
16. Özcan, E., Kheiri, A.: A hyper-heuristic based on random gradient, greedy and dominance. In: Gelenbe, E., Lent, R., Sakellari, G. (eds.) Computer and Information Sciences II, pp. 557–563. Springer, London (2012)

Energy-Efficiency Evaluation of Computation Offloading in Personal Computing

Yongpil Yoon, Georgia Sakellari$^{(\boxtimes)}$, Richard J. Anthony,
and Avgoustinos Filippoupolitis

Department of Computing and Information Systems,
University of Greeenwich, London, UK
{yongpil.yoon,g.sakellari,r.j.anthony,a.filippoupolitis}@gre.ac.uk

Abstract. Cloud computing has become common practice for a wide
variety of user communities. Yet, the energy efficiency and end-to-end
performance benefits of cloud computing are not fully understood. Here,
we focus specifically on the trade-off between local power saving and
increased execution time when work is offloaded from a user's PC to a
cloud environment. We have set up a 14-node private cloud and have executed a variety of applications with different processing demands. We have
measured the energy cost at the level of the individual user's PC, at the
level of the cloud, as well as at the two combined, contrasted to the execution time for each application when running on the PC and when running
on the cloud. Our results indicate that the tradeoff between energy cost
and performance differs considerably between applications of different
types. In most cases investigated, the total increase in energy consumption, incurred by running that additional application, was reduced significantly. This shows that research on using cloud computing as a means
to reduce the overall carbon footprint of IT is warranted. Of course, the
energy gains were more pronounced for energy-selfish users, who are only
interested in reducing their own carbon footprint, but these savings came
at the expense of performance, with execution time increase ranging from
1 % to 84 % for different applications.

Keywords: Cloud · Computation offloading · Energy · Performance ·
OpenStack

1 Introduction

Cloud computing has become a common paradigm for computational resource
provision. This paper investigates the viability of computation offloading to a
cloud for personal computers (PCs) with regard to reducing energy costs. In
other words, can computation offloading reduce the amount of required energy
for a PC to complete certain tasks? And what is the overall energy consumed
by the PC and the cloud in this case?

© The Author(s) 2016
T. Czachórski et al. (Eds.): ISCIS 2016, CCIS 659, pp. 163–171, 2016.
DOI: 10.1007/978-3-319-47217-1_18

2 Related Work

Computation offloading means executing certain tasks on more resourceful computers which are not in the user's immediate computing environment, so as to: (1) reduce energy consumption of the user's computing device, and/or (2) improve the performance of computation. Computation offloading first began and has been studied mainly for mobile devices [1–5] because of the noticeable difference in computation power between mobile devices and cloud servers [6]. Performance difference between PCs and computing resources from cloud providers is often negligible and sometimes PCs outperform cloud computing resources. Although resources from clouds can be massively scalable, it may not be cost-effective depending on factors such as the type of tasks to offload, required amount of data transmission, acceptable latency etc. [7,8]. Therefore, it is important to know under what circumstances offloading is beneficial for PCs.

For mobile devices, proposed techniques may differ slightly in architectures or implementations but all share the same fundamental idea, that a mobile device can stay idle or compute less by offloading parts of program code to the cloud. Most implementations, such as Phone2Cloud [9], Cuckoo [10], COMET [11] and MAUI [12], focus on identifying tasks that can be offloaded at runtime and how this can be achieved. Recently, other perspectives of computation offloading, such as energy consumption, have been investigated. For example, the energy cost of additional communication for offloading has been addressed in [13] in order to make more energy-efficient offloading decisions in cellular networks. Computation offloading as a service for mobile devices has been suggested by [14] to bridge the gaps between the offloading demands of mobile devices and the general computing resources, such as VMs, provided by commercial cloud providers. Energy-aware scheduling of the executions of offloaded computation into the cloud has been studied in [15].

3 Experimental Methodology

We have chosen to scope our initial investigation around the energy usage considered in isolation to provide an important baseline for further work, which will take into account additional aspects including the energy cost of network communication and the additional latency of the transfers. To evaluate whether computation offloading is beneficial for PCs in terms of power consumption, we have conducted experiments using a real world private cloud. In our experiments, computation is offloaded at the application level which means the entire execution of application software was offloaded to the cloud rather than offloading some parts of computation (function/method level) like existing offloading techniques for mobile devices, e.g., MAUI - method level (RPC-like) [12], Cuckoo - method level (RMI-like) [10]. Different applications which require different amounts of computation were run both locally on a PC and remotely on a VM created in our private cloud. In the case of offloading, the VM ran the application and sent the results back to the PC or saved resulting files in the cloud when completed.

The total execution time of each application was measured as well as the power loads (Wattage) of the PC and Cloud servers during this execution time, at one-second intervals.

The experiments were conducted on a Dell Optiplex 7010 desktop machine running the Linux operating system (Ubuntu 14.04.1). The PC has Intel Core i5-3550 3.30 GHz (Quad core), 16 GB DDR3 1600 MHz memory, and 750 GB SATA-II hard drive. The power-management configurations of the PC and the OS were not changed from their default settings, e.g., sleep, hibernate, disk spin-down configurations. It was possible that the screen timeout occurs in the PC while waiting for the completion of remote execution but the power consumed by its display (monitor) was not measured. Also, the applications executed in the cloud sent the current progress of computation back to the PC after the execution had finished.

Our cloud testbed was a private *OpenStack*[1] cloud infrastructure consisting of 14 machines, each with 4-core Intel Xeon E5-2407 2.20 GHz, 48 GB DDR3 1333 MHz ECC registered memory, and 500 GB SCSI hard drive. A virtual machine with 4 virtual cores (vCPU), 8 GB memory, and 40 GB disk space was used to run the offloaded computations. There was no background traffic in the cloud during our experiments. In order to measure the power consumption of the PC a *Watts up? .Net* energy meter[2] was used. It can measure wattage to the nearest tenth of a watt with an accuracy of ±1.5 %. The meter logged the power load of the PC at 1 s intervals during the executions.

In our experiments, the computation power used by a VM in the cloud is very similar to (but slightly lower than) the user PC's. If a more powerful VM is used, our results might be different. We plan to expand our experiments to investigate the effect that the different VM configurations and PC specifications have in both the introduced power consumption and the performance of each application. However, to put things into context, the VM used is considered quite large for cloud providers. For example, Microsoft Azure considers VM instances with 4 virtual cores and 7 GB RAM as large and VM instances with 8 cores and 14 GB RAM as extra large[3]. A more powerful VM than the one we used will cost considerably more to the PC user, neutralising at least any financial benefit of the corresponding energy savings. The cost of the PC user to access the cloud is an aspect that we do not take into account here, but will also consider in the next steps of our research.

We have chosen four different applications for our experiments with the primary criterion that they are computationally intensive. All four were executed with a multithreading/multiprocessing option apart from *SCID vs. PC* which runs only on a single core. *SCID vs. PC* is a chess toolkit, which requires continuous data transmission for drawing its graphical user interface when run remotely. We ran chess engine vs. chess engine tournament which requires computation

---

[1] http://www.openstack.org.

[2] https://www.wattsupmeters.com.

[3] https://azure.microsoft.com/en-us/documentation/articles/cloud-services-sizes-specs/.

for searching through databases. *avconv* is an open source video and audio converting program. It is a command line program and takes video or audio files as its input and writes converted files to the disk. Video transcoding involves heavy computation as well as constant read and write to a disk is required. A 1080 p 30 fps video file of 886 MB size encoded using x264 codec was used as input data and the video was converted to a h264 mp4 file. *pi_mp.py* is a multi-threaded python implementation of π estimation using Monte Carlo method. 200 million random points were used to estimate π in each execution. It requires repetitive arithmetic calculations and a large amount of memory. *Blender* is op, featuring 3D modelling, video editing, camera object tracking, etc. In our experiments, a demo file provided by blender, called BMW benchmark, was rendered from command line. The output of the rendering is a JPEG file.

The results of the executions were sent back to the PC if it was simply text output, but if an application needed to write a file, that was saved in the cloud (in the VM where the application was executed) and thus the execution time we measured in the latter case does not include the transmission time of the resulting files. Neither the PC nor the VM in the cloud performed any other user-level activity during our experiment. There is some natural variance in the power usage of the cloud infrastructure, comprising as it does 8 compute nodes in a rack, subject to temperature fluctuations. We have found that this variation was in the worst case 3.2 %. To reduce the impact of noise in the measured cloud power usage, each application run was repeated 10 times and the average values are used in the results presented here.

4   Experimental Results on Power Consumption Vs. Performance Tradeoff

To investigate the effect of computation offloading on the energy consumption of PCs, we focus on the nature of the tradeoff between power consumption and performance. For the latter, we use the total execution time for each application, measured experimentally when running locally and when offloaded to the cloud. We have also calculated the energy consumption of the PC and the cloud (power consumption × execution time) during the executions.

4.1   Power Consumption and Performance

First we established a baseline power consumption for the PC and likewise for the cloud. The cloud required 1036.00 W on average when IDLE while the PC required only 22.23 W when IDLE. The cloud requires much more power compared to the PC since it has more machines which are power-hungrier than the PC. Obviously, the PC requires noticeably less power while simply waiting for the cloud to finish the execution, than when running applications locally. The part a's (left column) of Figs. 1, 2, 3 and 4 show the execution time vs. power consumption tradeoff. When only one core is used, about 40 % less power is required ("SCID vs PC") and when four cores are used, nearly 70 % less power is required

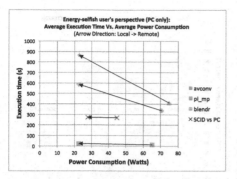
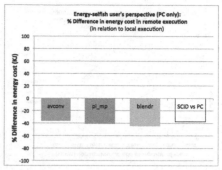

(a) (a) Average Execution Time Vs. Average Power Consumption

(b) (b) % Difference in PC energy cost in remote execution (in relation to local execution)

Fig. 1. Energy-selfish user's perspective (PC only)

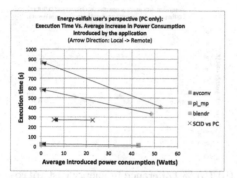
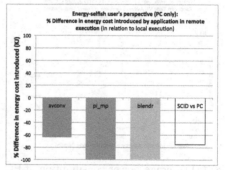

(a) (a) Execution Time Vs. Average **Increase** in Power Consumption introduced by the application

(b) (b) % Difference in PC energy cost **introduced** by an application in remote execution (in relation to local execution)

Fig. 2. Energy-selfish user's perspective (PC only): **Increases** introduced by the applications in remote operation

on average, but if seen in isolation, this is misleading. The average power load only represents the power consumption per unit time and thus, the total amount of energy consumed by each application depends on the execution time, as seen in the part b's (right column) of Figs. 1, 2, 3 and 4. The cloud required 1054.47 W of power on average during the executions. However, the introduced power load by the executions of the PC (the difference between the average power load when applications are running and when IDLE) was 41.63 W on average, while the average introduced power load in the cloud was only 18.47 W. When computation was offloaded almost all applications took much longer (up to 84 % longer) to finish certain tasks, although the VM in the cloud has the same number of

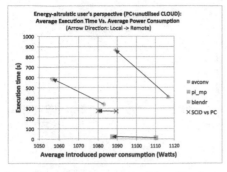
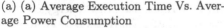
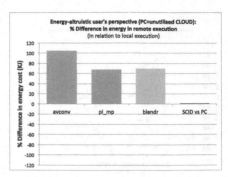

(a) (a) Average Execution Time Vs. Average Power Consumption

(b) (b) % Difference in total energy cost in remote execution (in relation to local execution)

Fig. 3. Energy-altruistic user's perspective (PC+CLOUD)

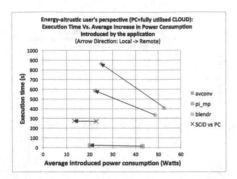
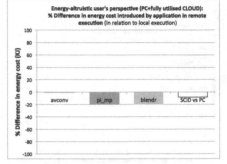

(a) (a) Average power consumption of PC and Cloud **introduced** by each application

(b) (b) % Difference in energy cost **introduced** by an application in remote execution (in relation to local execution)

Fig. 4. Energy-altruistic user's perspective (PC+CLOUD): **Increases** introduced by the applications in remote operation

processors as the PC. The additional end-to-end time includes network transfer latency, but this was very low because of the small amount of data needed to be transmitted. Any execution time increases were mainly due to the lower computing power of the VM in the cloud (vCPUs vs. real CPUs). Although less power is required per unit time when computation is offloaded, the total amount of energy required increases in proportion to the execution time.

4.2 Energy Consumption

The part b's (right column) of Figs. 1, 2, 3 and 4 show the percentage of the energy difference consumed on average by each application over 10 runs each, both from the PC user perspective and the total (PC+cloud) perspective.

Based on our results, the energy Vs. performance tradeoff introduced by computation offloading differs considerably depending on the application and on the perspective taken. We can broadly classify energy-conscious users as either "energy-selfish users", who are interested only in reducing the energy cost of their own PCs, versus "energy-altruistic users", who are interested in the overall reduction of the energy cost of their computation, which includes both their PC and the Cloud infrastructure. For the sake of simplicity, we have not considered energy costs introduced by the network connection to the cloud. The two terms may make sense from a societal angle where human users may be interested in reducing their own devices' energy consumption only or may care about reducing the total environmental impact of their computation, but they can also have practical technical meaning from a system perspective. For instance, an energy-selfish entity could be a battery-operated device, such as a vehicle, a wearable device or a sensor, which for operational reasons is designed to offload its computation to a cloud infrastructure that is not resource-constrained.

For energy-selfish users, we have observed that offloading is most beneficial for the application that runs on a single core, as the local power consumption dropped significantly without a noticeable increase in execution time. The other three applications also experienced considerable reduction in local power consumption, but mostly at a noticeable expense in execution time. Overall, all applications have considerably reduced local energy usage when offloaded (varying from 63.75 % up to 98.88 % reduction in energy introduced by the application compared to local execution). For energy-altruistic users, we have also observed that offloading clearly benefits the single-core application, since, again, the execution time does not increase much, but for the rest of the applications executing them remotely significantly increases the total energy of the system, simply because the energy costs for running a cloud are much higher. Looking at the applications in isolation though, the total amount of energy introduced by each one is less for remote execution (varying from 0.97 % up to 20.28 %) compared to local execution.

5   Conclusions and Future Work

This paper has studied the viability of computation offloading for PCs with respect to the energy Vs. performance tradeoff for computationally heavy applications. We see that in most cases, the user can sacrifice performance to make considerable energy savings, not only locally, but also when the total energy cost, including the cloud's, is taken into account. If a cloud infrastructure already exists and runs applications, adding one more incurs less total energy cost at the PC and cloud than a new application would incur running on the PC only. This is significant because it shows that adopting cloud computing can be a meaningful option for reducing the overall carbon footprint of IT. For energy-selfish users, only interested in reducing their own carbon footprint, these savings are considerably greater. In both cases, the energy savings come at the expense of performance. In our experiments, the execution time increase ranged between

1 % and 84 % depending on the application. These initial experiments have provided a valuable baseline for exploration and we plan to extend them for different VM configurations. Looking at other areas of future work, we will investigate simultaneous executions of many computationally light applications. This will yield more accurate relation between the amount of energy saved and other factors like computation power of the cloud and the heaviness of applications that are offloaded.

References

1. Kumar, K., Liu, J., Lu, Y.H., Bhargava, B.: A survey of computation offloading for mobile systems. Mob. Netw. Appl. **18**(1), 129–140 (2013)
2. Gelenbe, E., Lent, R.: Energy-QoS trade-offs in mobile service selection. Future Internet **5**(2), 128–139 (2013)
3. Rahimi, M.R., Ren, J., Liu, C.H., Vasilakos, A.V., Venkatasubramanian, N.: Mobile cloud computing: a survey, state of art and future directions. Mob. Netw. Appl. **19**(2), 133–143 (2014)
4. Othman, M., Madani, S.A., Khan, S.U.: A survey of mobile cloud computing application models. IEEE Commun. Surv. Tutorials **16**(1), 393–413 (2014)
5. Gelenbe, E., Lent, R.: Optimising server energy consumption and response time. Theor. Appl. Inform. **24**(4), 257–270 (2012)
6. Sakellari, G., Loukas, G.: A survey of mathematical models, simulation approaches and testbeds used for research in cloud computing. Simul. Modell. Pract. Theor. **39**, 92–103 (2013)
7. Kumar, K., Lu, Y.H.: Cloud computing for mobile users: can offloading computation save energy? Computer **43**(4), 51–56 (2010)
8. Gelenbe, E., Lent, R., Douratsos, M.: Choosing a local or remote cloud. In: Network Cloud Computing and Applications, pp. 25–30 (2012)
9. Xia, F., Ding, F., Li, J., Kong, X., Yang, L.T., Ma, J.: Phone2Cloud: exploiting computation offloading for energy saving on smartphones in mobile cloud computing. Inf. Syst. Front. **16**(1), 95–111 (2014)
10. Kemp, R., Palmer, N., Kielmann, T., Bal, H.: Cuckoo: a computation offloading framework for smartphones. In: Gris, M., Yang, G. (eds.) MobiCASE 2010. LNICSSITE, vol. 76, pp. 59–79. Springer, Heidelberg (2012). doi:10.1007/978-3-642-29336-8_4
11. Gordon, M.S., Jamshidi, D.A., Mahlke, S., Mao, Z.M., Chen, X.: COMET: code offload by migrating execution transparently. In: Proceedings of the USENIX Symposium on Operating Systems Design and Implementation, pp. 93–106 (2012)

12. Cuervo, E., Balasubramanian, A., Cho, D.K., Wolman, A., Saroiu, S., Chandra, R., Bahl, P.: MAUI: making smartphones last longer with code offload. In: Proceedings of ACM Mobile systems, applications, and services, pp. 49–62 (2010)
13. Geng, Y., Hu, W., Yang, Y., Gao, W., Cao, G.: Energy-efficient computation offloading in cellular networks. In: IEEE ICNP, pp. 145–155 (2015)
14. Shi, C., Habak, K., Pandurangan, P., Ammar, M., Naik, M., Zegura, E.: Cosmos: computation offloading as a service for mobile devices. In: Proceedings of ACM MobiHoc, pp. 287–296 (2014)
15. Zhang, W., Wen, Y., Wu, D.O.: Energy-efficient scheduling policy for collaborative execution in mobile cloud computing. In: IEEE INFOCOM, pp. 190–194 (2013)

Queuing Systems

Analysis of Transient Virtual Delay in a Finite-Buffer Queueing Model with Generally Distributed Setup Times

Wojciech M. Kempa[2] and Dariusz Kurzyk[1,2($\boxtimes$)]

[1] Institute of Theoretical and Applied Informatics, Polish Academy of Sciences,
Bałtycka 5, 44-100 Gliwice, Poland
dkurzyk@iitis.pl
[2] Institute of Mathematics, Silesian University of Technology,
Kaszubska 23, 44-100 Gliwice, Poland

Abstract. Time-dependent queueing delay (virtual waiting time) distribution conditioned by the initial level of buffer saturation is considered in a finite model with Poisson arrivals, generally distributed service times and setup times preceding the first processing in each busy period. Applying theoretical approach based on the idea of embedded Markov chain, integral equations and some results from linear algebra, a compact-form representation for the Laplace transform of queueing delay distribution is obtained. Analytical results are illustrated via numerical considerations confirmed by process-based discrete-event simulations.

1 Introduction

Queueing systems with different types of restrictions in access to the service station (server) are being intensively studied nowadays, in view of their use in modeling many phenomena occurring in technical sciences and economics. Particularly important here are models with a limited maximal number of customers (packets, calls, jobs, etc.), which naturally can describe systems with losses due to buffer overflows (buffers of input/output interfaces in TCP/IP routers, accumulating buffers in production systems). In many practical systems, which can be described by queueing models, a mechanism of turning off the server at the time when the system becomes empty is implemented; the server is being activated when the first customer arrives after the period of inactivity. The use of such a mechanism is often being forced to save energy that the server uses to remain on standby despite the lack of applications in the system (wireless networks, automated production lines, etc.). It happens quite often that the waking up of service station (restart) is not simultaneous with the start of processing in "normal" mode. The server may indeed need some time (usually random) to achieve full readiness to work. Assuming randomness of setup time, such a mechanism could be called probabilistic waking up the server. For example, a node of wireless network working under the Wi-Fi standard (IEEE 802.11) wakes thereby regularly just before sending the beacon frame from the access point [7,8].

© The Author(s) 2016
T. Czachórski et al. (Eds.): ISCIS 2016, CCIS 659, pp. 175–184, 2016.
DOI: 10.1007/978-3-319-47217-1_19

In [6] $M/G/1$-type queuing system with server vacations and setup times is used to model sleeping mode in cellular network. A similar phenomenon can also be observed e.g. in production lines: after restarting, a machine needs a certain, often random, time to achieve its full readiness to work. Furthermore, the formula relating with waiting time in stationary state of $GI/G/1$-type queues with setup times can be found in [2,3].

2   Mathematical Model

In this section we state mathematical description of the considered queueing model and introduce necessary notation and definitions. So, we deal with the finite $M/G/1/K$-type model in which packets (calls, jobs, customers, etc.) arrive according to a Poisson process with rate λ and are processed individually, basing on the FIFO service discipline, with a CDF (=cumulative distribution function) $F(\cdot)$. The system capacity is bounded by a non-random value K, i.e. we have a finite buffer with $K-1$ places and one place reserved for service. Every time when the system becomes empty the server is being switched off (an idle period begins). Simultaneously with the arrival epoch of the packet incoming into the empty system, a server setup time begins, which is generally distributed random variable with a CDF $G(\cdot)$. The setup time is needed for the server to reach full ability for job processing, hence during setup times the service process is suspended. Let $f(\cdot)$ and $g(\cdot)$ be LSTs (=Laplace-Stieltjes transforms) of CDFs $F(\cdot)$ and $G(\cdot)$, respectively, i.e. for $\text{Re}(s) > 0$

$$f(s) \stackrel{def}{=} \int_0^\infty e^{-st} dF(t), \quad g(s) \stackrel{def}{=} \int_0^\infty e^{-st} dG(t). \tag{1}$$

Let us denote by $X(t)$ the number of packets present in the system at time t (including the one being processed, if any) and by $v(t)$ the queueing delay (virtual waiting time) at time t, i.e. the time needed for the server to process all packets present at time t or, in other words, waiting time of hypothetical (virtual) packet arriving exactly at time t. Introduce the following notation:

$$V_n(t, x) \stackrel{def}{=} \mathbf{P}\{v(t) > x \,|\, X(0) = n\}dt, \quad t, x > 0, 0 \le n \le K, \tag{2}$$

for the transient queueing delay (tail) distribution, conditioned by the initial level of buffer saturation. We are interested in the explicit formula for the LT (=Laplace transform) of $V_n(t, x)$ in terms of "input" characteristics of the system, namely arrival rate λ, system capacity K, and transforms $f(\cdot)$ and $g(\cdot)$ of service and setup time distributions. We end this section with some additional notation which will be used throughout the paper. So, let

$$F^{0*}(t) = 1, \quad F^{k*}(t) = \int_0^t F^{(k-1)*}(t-y)dF(y), \quad k \ge 1, t > 0, \tag{3}$$

and introduce the notation $\overline{H}(t) \stackrel{def}{=} 1 - H(t)$, where $H(\cdot)$ is an arbitrary CDF. Moreover, let $I\{\mathbb{A}\}$ be the indicator of random event $\mathbb{A}$.

3   Integral Equations for Transient Queueing Delay Distribution

In this section, by using the paradigm of embedded Markov chain and the formula of total probability we build the system of equations for conditional time-dependent virtual delay distribution defined in (2). Next, we build the system for Laplace transforms corresponding to the original one.

Assume, firstly, that the system is empty before the opening, so its evolution begins with idle period and the setup time begins simultaneously with the arrival epoch of the first batch of packets. We can, in fact, distinguish three mutually exclusive random events:

(1) the first arrival occurs before t and the setup time also completes before t (we denote this event by $E_1(t)$);
(2) the first packet (call, job, customer, etc.) arrives before t but the setup time completes after t $(E_2(t))$;
(3) the first arrival occurs after time t $(E_3(t))$.

Let us define

$$V_0^{(i)}(t, x) \overset{def}{=} \mathbf{P}\{(v(t) > x) \cap E_i(t) \,|\, X(0) = 0\}, \tag{4}$$

where $t, x > 0$, $0 \leq m \leq K$ and $i = 1, 2, 3$. So, for example, $V_0^{(3)}(t, x)$ denotes the probability that queueing delay at time t exceeds x and the first arrival occurs after t, on condition that the system is empty at the opening (at time $t = 0$). Obviously, we have

$$V_0(t, x) = \mathbf{P}\{v(t) > x \,|\, X(0) = 0\} = \sum_{i=1}^{3} V_0^{(i)}(t, x) \tag{5}$$

Let us note that the following representation is true:

$$V_0^{(1)}(t, x) = \int_{y=0}^{t} \lambda e^{-\lambda y} dy \int_{u=0}^{t-y} \left[\sum_{i=0}^{K-2} \frac{(\lambda u)^i}{i!} e^{-\lambda u} V_{i+1}(t - y - u, x) \right.$$
$$\left. + V_K(t - y - u, x) \sum_{i=K-1}^{\infty} \frac{(\lambda u)^i}{i!} e^{-\lambda u} \right] dG(y). \tag{6}$$

Let us comment on (6) briefly. Indeed, the first summand on the right side describes the situation in which the buffer does not become saturated during the setup time, while the second one relates to the case in which a buffer overflow occurs during the setup time. Similarly, taking into consideration the random event E_2, we find

$$V_0^{(2)}(t, x) = \int_{y=0}^{t} \lambda e^{-\lambda y} \int_{u=t-y}^{\infty} \sum_{i=0}^{K-2} \frac{[\lambda(t-y)]^i}{i!} e^{-\lambda(t-y)} \overline{F}^{(i+1)*}(x - y - u + t) dG(u) dy. \tag{7}$$

Finally we have, obviously,

$$V_0^{(3)}(t, x) = 0. \tag{8}$$

Referring to (5), we obtain from (6)–(8)

$$
V_0(t, x) = \int_{y=0}^t \lambda e^{-\lambda y} dy \int_{u=0}^{t-y} \left[\sum_{i=0}^{K-2} \frac{(\lambda u)^i}{i!} e^{-\lambda u} V_{i+1}(t-y-u, x) \right.
$$

$$
\left. + V_K(t-y-u, x) \sum_{i=K-1}^{\infty} \frac{(\lambda u)^i}{i!} e^{-\lambda u} \right] dG(y)
$$

$$
+ \int_{y=0}^t \lambda e^{-\lambda y} \int_{u=t-y}^{\infty} \sum_{i=0}^{K-2} \frac{[\lambda(t-y)]^i}{i!} e^{-\lambda(t-y)} \overline{F}^{(i+1)*}(x-y-u+t) dG(u) dy. \tag{9}
$$

Now, let us take into consideration the situation in which the system is not empty primarily (at time $t = 0$), i.e. $1 \leq n \leq K$. Due to the fact that successive departure moments are Markov times in the evolution of the $M/G/1$-type system (see e.g. [1]), then, applying the continuous version of Total Probability Law with respect to the first departure moment after $t = 0$, we get the following system of integral equations:

$$
V_n(t, x) = \int_0^t \left[\sum_{i=0}^{K-n-1} \frac{(\lambda y)^i}{i!} e^{-\lambda y} V_{n+i-1}(t-y, x) + V_{K-1}(t-y, x) \sum_{i=K-n}^{\infty} \frac{(\lambda y)^i}{i!} e^{-\lambda y} \right] dF(y)
$$

$$
+ I\{1 \leq n \leq K-1\} \sum_{i=0}^{K-n-1} \frac{(\lambda t)^i}{i!} e^{-\lambda t} \int_t^{\infty} \overline{F}^{(n+i-1)*}(x-y+t) dF(y), \tag{10}
$$

where $1 \leq n \leq K$. The interpretation of the first two summands on the right side of (10) is similar to (6)-(7). The last summand on the right side relates to the situation in which the first service completion epoch occurs after time t; in such a case, if $n = K$, the queueing delay at time t equals 0, since the "virtual" packet arriving at this time is lost because of the buffer overflow. Let us introduce the following notation:

$$
\widehat{v}_n(s, x) \overset{def}{=} \int_0^{\infty} e^{-st} V_n(t, x) dt, \quad \text{Re}(s) > 0, 0 \leq n \leq K. \tag{11}
$$

where $\text{Re}(s) > 0$ and $0 \leq n \leq K$. By the fact that for $\text{Re}(s) > 0$ we have

$$
\int_{t=0}^{\infty} e^{-st} dt \int_{y=0}^t \lambda e^{-\lambda y} dy \int_{u=0}^{t-y} \frac{(\lambda u)^i}{i!} e^{-\lambda u} V_j(t-y-u, x) dG(u)
$$

$$
= \int_{y=0}^{\infty} \lambda e^{-(\lambda+s)y} dy \int_{u=0}^{\infty} e^{-(\lambda+s)u} \frac{(\lambda u)^i}{i!} e^{-\lambda u} dG(u) \int_{t=y+u}^{\infty} e^{-s(t-y-u)}
$$

$$
\times V_j(t-y-u, x) dt = a_i(s) \widehat{v}_j(s, x), \tag{12}
$$

where

$$a_i(s) \stackrel{def}{=} \frac{\lambda}{\lambda+s} \int_0^\infty \frac{(\lambda y)^i}{i!} e^{-(\lambda+s)y} dG(y), \tag{13}$$

we obtain from (9)

$$\widehat{v}_0(s,x) = \sum_{i=0}^{K-2} a_i(s)\widehat{v}_{i+1}(s,x) + \widehat{v}_K(s,x) \sum_{i=K-1}^{\infty} a_i(s) + \eta(s,x), \tag{14}$$

where we denote

$$\eta(s,x) \stackrel{def}{=} \int_0^\infty e^{-st} V_0^{(2)}(t,x) dt$$

$$= \int_{t=0}^{\infty} e^{-(s+\lambda)t} dt \int_{y=0}^{t} \sum_{i=0}^{K-2} \frac{\lambda^{i+1}(t-y)^i}{i!} dy \int_{u=t-y}^{\infty} \overline{F}^{(i+1)*}(x-y-u+t) dG(u). \tag{15}$$

Similarly, denoting

$$\alpha_i(s) \stackrel{def}{=} \int_0^\infty e^{-(\lambda+s)x} \frac{(\lambda x)^i}{i!} dF(x) \tag{16}$$

and

$$\kappa_n(s,x) \stackrel{def}{=} I\{1 \le n \le K-1\} \int_{t=0}^{\infty} \sum_{i=0}^{K-n-1} e^{-(s+\lambda)t} \frac{(\lambda t)^i}{i!} \int_t^\infty \overline{F}^{(n+i-1)*}(x-y+t) dF(y) dt, \tag{17}$$

where $\text{Re}(s) > 0$, we transform the equations (10) as follows:

$$\widehat{v}_n(s,x) = \sum_{i=0}^{K-n-1} \alpha_i(s)\widehat{v}_{n+i-1}(s,x) + \widehat{v}_{K-1}(s,x) \sum_{i=K-n}^{\infty} \alpha_i(s) + \kappa_n(s,x), \tag{18}$$

where $1 \le n \le K$. Let us define

$$z_n(s,x) \stackrel{def}{=} \widehat{v}_{K-n}(s,x), \quad 0 \le n \le K. \tag{19}$$

After introducing (19), we obtain from (18) the following equations:

$$\sum_{i=-1}^{n} \alpha_{i+1}(s) z_{n-i}(s,x) - z_n(s,x) = \psi_n(s,x), \tag{20}$$

where $0 \le n \le K-1$, and the sequence $\psi_n(s,x)$ is defined as follows:

$$\psi_n(s,x) \stackrel{def}{=} \alpha_{n+1}(s) z_0(s,x) - z_1(s,x) \sum_{i=n+1}^{\infty} \alpha_i(s) - \kappa_{K-n}(s,x). \tag{21}$$

Similarly, utilizing (19) in (14), we get

$$z_K(s,x) = \sum_{i=0}^{K-2} a_i(s)z_{K-i-1}(s,x) + z_0(s,x) \sum_{i=K-1}^{\infty} a_i(s) + \eta(s,x). \qquad (22)$$

In the next section we obtain a compact-form solution of the system (20) and (22) written in terms of "input" system characteristics and a certain functional sequence defined recursively by coefficients $\alpha_i(s)$, $i \geq 0$.

4   Compact Solution for Queueing Delay Transforms

In [4] (see also [5]) the following linear system of equations is investigated:

$$\sum_{i=-1}^{n} \alpha_{i+1}z_{n-i} - z_n = \psi_n, \quad n \geq 0, \qquad (23)$$

where z_n, $n \geq 0$, is a sequence of unknowns and α_n and ψ_n, $n \geq 0$, are known coefficients, where $\alpha_0 \neq 0$. It was proved (see [4]) that each solution of (23) can be written in the following way:

$$z_n = CR_{n+1} + \sum_{i=0}^{n} R_{n-i}\psi_i, \quad n \geq 0, \qquad (24)$$

where C is a constant and terms of the sequence (R_n), $n \geq 0$, can be computed in terms of α_n, $n \geq 0$, recursively in the following way:

$$R_0 = 0, \quad R_1 = \alpha_0^{-1}, \quad R_{n+1} = R_1\left(R_n - \sum_{i=0}^{n} \alpha_{i+1}R_{n-i}\right), \quad n \geq 1. \qquad (25)$$

Observe that the system (20) has the same form as (23) but with coefficients α_i and ψ_i, $i \geq 0$, depending on s and (s,x), respectively. Thus, the solution of (20) can be derived by using (24). The fact that the number of equations in (24) (comparing to (20)) is finite, allows for finding $C = C(s,x)$ in the explicit form, treating the equation (22) as a boundary condition. Hence, we obtain the following formula (see (23)–(25)):

$$z_n(s,x) = C(s,x)R_{n+1}(s) + \sum_{i=0}^{n} R_{n-i}(s)\psi_i(s,x), \quad n \geq 0, \qquad (26)$$

where the functional sequence $(R_n(s))$, $n \geq 0$, is defined by

$$R_0(s)=0, \quad R_1(s)=\alpha_0^{-1}(s), \quad R_{n+1}(s)=R_1(s)\left(R_n(s)-\sum_{i=0}^{n}\alpha_{i+1}(s)R_{n-i}(s)\right), \qquad (27)$$

where $n \geq 1$ and $\alpha_i(s)$ is stated in (16). Taking $n = 0$ in (26), we obtain the following representation:

$$z_0(s, x) = C(s, x)R_1(s) \tag{28}$$

and substituting $n = 1$, we get

$$z_1(s, x) = C(s, x)R_2(s) + R_1(s)\psi_0(s, x)$$

$$= C(s, x)R_2(s) + R_1(s)\Big(\alpha_1(s)R_1(s)C(s, x) - z_1(s, x)\sum_{i=1}^{\infty}\alpha_i(s)\Big), \tag{29}$$

since $\kappa_K(s, x) = 0$. From (29) we obtain

$$z_1(s, x) = \theta(s)C(s, x)\big(R_2(s) + \alpha_1(s)R_1^2(s)\big), \tag{30}$$

where

$$\theta(s) \stackrel{def}{=} \Big[1 + R_1(s)\sum_{i=1}^{\infty}\alpha_i(s)\Big]^{-1} = \frac{f(\lambda + s)}{f(s)}. \tag{31}$$

Now the formulae (28) and (30)–(31) allows for writing terms of the functional sequence $(\psi_n(s, x))$, $n \geq 0$ (see (21)), as a function of $C(s, x)$. In order to find the representation for $C(s, x)$, we must rewrite the formula (22), utilizing identities (21), (26), (28) and (30). We obtain

$$z_K(s, x) = \sum_{i=1}^{K-1} a_{K-i-1}(s)\Big[C(s, x)R_{i+1}(s) + \sum_{j=0}^{i} R_{i-j}(s)\psi_j(s, x)\Big]$$

$$+ C(s, x)R_1(s)\sum_{i=K-1}^{\infty} a_i(s) + \eta(s, x) = \sum_{i=1}^{K-1} a_{K-i-1}(s)\Big[C(s, x)R_{i+1}(s)$$

$$+ \sum_{j=0}^{i} R_{i-j}(s)\Big(\alpha_{j+1}(s)z_0(s, x) - z_1(s, x)\sum_{r=j+1}^{\infty}\alpha_r(s) - \kappa_{K-j}(s, m)\Big)\Big]$$

$$+ C(s, x)R_1(s)\sum_{i=K-1}^{\infty} a_i(s) + \eta(s, x) = C(s, x)\Big\{\sum_{i=1}^{K-1} a_{K-i-1}(s)\Big[R_{i+1}(s) + \sum_{j=0}^{i} R_{i-j}(s)$$

$$\times \Big(R_1(s)\alpha_{j+1}(s) - \theta(s)\big(R_2(s) + \alpha_1(s)R_1^2(s)\big)\sum_{r=j+1}^{\infty}\alpha_r(s)\Big)\Big] + R_1(s)\sum_{i=K-1}^{\infty} a_i(s)\Big\}$$

$$- \sum_{i=1}^{K-1} a_{K-i-1}\sum_{j=1}^{i} R_{i-j}(s)\kappa_{K-j}(s, x) + \eta(s, x) = \Phi_1(s)C(s, x) + \chi_1(s, x), \tag{32}$$

where we denote

$$\Psi_1(s) \stackrel{def}{=} \sum_{i=1}^{K-1} a_{K-i-1}(s)\Big[R_{i+1}(s) + \sum_{j=0}^{i} R_{i-j}(s)\Big(R_1(s)\alpha_{j+1}(s)$$

$$- \theta(s)\big(R_2(s) + \alpha_1(s)R_1^2(s)\big)\sum_{r=j+1}^{\infty}\alpha_r(s)\Big] + R_1(s)\sum_{i=K-1}^{\infty} a_i(s) \tag{33}$$

and

$$\chi_1(s,x) \stackrel{def}{=} -\sum_{i=1}^{K-1} a_{K-i-1} \sum_{j=1}^{i} R_{i-j}(s)\kappa_{K-j}(s,x) + \eta(s,x). \tag{34}$$

Finally, let us substitute $n = K$ in (26) and apply the formulae (21), (28) and (30). We get

$$\begin{aligned}
z_K(s,x) &= C(s,x)R_{K+1}(s) + \sum_{i=0}^{K} R_{K-i}(s)\Big\{\alpha_{i+1}(s)R_1(s)C(s,x) \\
&\quad - \theta(s)C(s,x)\big(R_2(s)+\alpha_1(s)R_1^2(s)\big)\sum_{j=i+1}^{\infty}\alpha_j(s) - \kappa_{K-i}(s,x)\Big\} \\
&= C(s,x)\Big\{R_{K+1}(s) + \sum_{i=0}^{K} R_{K-i}(s)\big[\alpha_{i+1}(s)R_1(s) - \theta(s)\big(R_2(s)+\alpha_1(s)R_1^2(s)\big) \\
&\quad \times \sum_{j=i+1}^{\infty}\alpha_j(s)\big]\Big\} - \sum_{i=1}^{K} R_{K-i}(s)\kappa_{K-i}(s,x)\Big) = \Psi_2(s)C(s,x) + \chi_2(s,x), \tag{35}
\end{aligned}$$

where

$$\Psi_2(s) \stackrel{def}{=} R_{K+1}(s) + \sum_{i=0}^{K} R_{K-i}(s)\Big[\alpha_{i+1}(s)R_1(s) - \theta(s)\big(R_2(s)+\alpha_1(s)R_1^2(s)\big)\sum_{j=i+1}^{\infty}\alpha_j(s)\Big] \tag{36}$$

and

$$\chi_2(s,x) \stackrel{def}{=} -\sum_{i=1}^{K} R_{K-i}(s)\kappa_{K-i}(s,x). \tag{37}$$

Comparing the right sides of (32) and (35), we eliminate $C(s,x)$ as follows:

$$C(s,x) = \big[\Psi_1(s) - \Psi_2(s)\big]^{-1}\big[\chi_2(s,x) - \chi_1(s,x)\big]. \tag{38}$$

Now, from the formulae (19), (26) and (38), we obtain the following main result:

Theorem 1. *The representation for the LT of the conditional transient queueing delay distribution in the $M/G/1/K$-type model with generally distributed setup times is the following:*

$$\begin{aligned}
\widehat{v}_n(s,x) &= \int_0^{\infty} e^{-st}\mathbf{P}\{v(t) > x \mid X(0) = n\}dt = \frac{\chi_2(s,x) - \chi_1(s,x)}{\Psi_1(s) - \Psi_2(s)}\Big\{R_{K-n+1}(s) \\
&\quad + \sum_{i=0}^{K-n} R_{K-n-i}(s)\big[\alpha_{i+1}(s)R_1(s) - \theta(s)\big(R_2(s)+\alpha_1(s)R_1^2(s)\big)\sum_{j=i+1}^{\infty}\alpha_j(s)\big]\Big\} \\
&\quad - \sum_{i=0}^{K-n} R_{K-n-i}(s)\kappa_{K-i}(s,m), \tag{39}
\end{aligned}$$

where the formulae for $\alpha_i(s)$, $\kappa_i(s,x)$, $R_i(s)$, $\theta(s)$, $\Psi_1(s)$, $\chi_1(s,x)$, $\Psi_2(s)$ and $\chi_2(s,x)$ are given in (16), (17), (27), (31), (33), (34), (36) and (37), respectively.

5   Numerical Example

Let us take into consideration a node of the wireless sensor network with buffer of size 6 packets, with the stream of packets of average size $100\,B$ arriving to the node according to a Poisson process with intensity $300\,Kb/s$. Hence, the $\lambda = 375$ packets per second arrive to the node and interarrival time between successive packets is equal to $2,7\,ms$. Subsequently, assume, that packets are being transmitted with speed $400\,Kb/s$ according to a 2-Erlang distribution with parameter $\mu = 1000$, that gives the mean processing time $2\,ms$. Moreover, let us consider that the radio transmitter of the node is switched off during an idle period and needs an exponentially distributed setup time to become ready for processing. Consider cases in which the mean times are equal to 1, 10, and $100\,ms$, respectively. The probabilities of $\mathbf{P}\{v(t) > x|X(0) = 0\}$ for $x = 0.001$ and $x = 0.01$ are presented in Fig. 1. The figures show that the analytical results are compatible with process-based discrete-event simulations (DES).

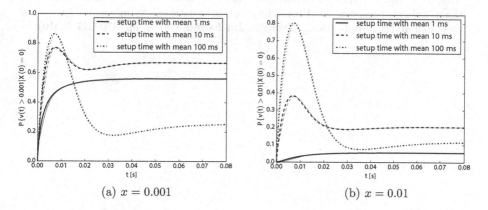

(a) $x = 0.001$ \qquad\qquad (b) $x = 0.01$

Fig. 1. Probabilities $\mathbf{P}\{v(t) > x|X(0) = 0\}$ for $x = 0.001$ (a) and $x = 0.01$ (b), where mean setup time is equal to 1 (solid line), 10 (dashed line) and 100 (dot dashed line) ms. Bold black lines and thin green lines correspond with analytical and DES results, respectively (Color figure online)

References

1. Cohen, J.W.: The Single Server Queue. Elsevier, Amsterdam (2012)
2. Gelenbe, E., Iasnogorodski, R.: A queue with server of walking type (autonomous service). Ann. de l'IHP Probabilités et Stat. **16**(1), 63–73 (1980)
3. Gelenbe, E., Mitrani, I.: Analysis and Synthesis of Computer Systems, vol. 4. World Scientific, Singapore (2010)
4. Korolyuk, V.S.: Boundary-Value Problems for Compound Poisson Processes. Naukova Dumka, Kiev (1975). (in Russian)
5. Korolyuk, V.S., Bratiichuk, N.S., Pirdzhanov, B.: Boundary-Value Problems for Random Walks. Ylym, Ashkhabad (1987)
6. Niu, Z., Guo, X., Zhou, S., Kumar, P.R.: Characterizing energy-delay tradeoff in hyper-cellular networks with base station sleeping control. IEEE J. Sel. Areas Commun. **33**(4), 641–650 (2015)
7. Sun, Q., Jin, S., Chen, C.: Energy analysis of sensor nodes in WSN based on discrete-time queueing model with a setup. In: Proceedings of 2010 Chinese Control and Decision Conference (CCDC), pp. 4114–4118. IEEE (2010)
8. Yue, W., Sun, Q., Jin, S.: Performance analysis of sensor nodes in a WSN with sleep/wakeup protocol. In: Proceedings of International Symposium Operations Research and its Applications, Chengdu-Jiuzhaigou, China, pp. 370–377 (2010)

Delays in IP Routers, a Markov Model

Tadeusz Czachórski[1]([✉]), Adam Domański[2], Joanna Domańska[1],
Michele Pagano[3], and Artur Rataj[1]

[1] Institute of Theoretical and Applied Informatics, Polish Academy of Sciences,
Baltycka 5, 44-100 Gliwice, Poland
tadek@iitis.pl
[2] Institute of Informatics, Silesian Technical University,
Akademicka 16, 44-100 Gliwice, Poland
adamd@polsl.pl
[3] Department of Information Engineering, University of Pisa,
Via Caruso 16, 56122 Pisa, Italy
m.pagano@iet.unipi.it

Abstract. Delays in routers are an important component of end-to-end
delay and therefore have a significant impact on quality of service. While
the other component, the propagation time, is easy to predict as the
distance divided by the speed of light inside the link, the queueing delays
of packets inside routers depend on the current, usually dynamically
changing congestion and on the stochastic features of the flows. We use
a Markov model taking into account the distribution of the size of packets
and self-similarity of incoming flows to investigate their impact on the
queueing delays and their dynamics.

Keywords: Markov queueing models · Self-similarity · IP packets
length distribution · IP routers delays

1 Introduction

Queueing theory has its origins in models proposed by Erlang and Engset a
hundred years ago for evaluation of telephone and telegraph systems. These
models were based on Markov chains, which since then accompany modelling
and evaluation of telecommunication systems. With the increasing complexity
of models they encounter natural limitations as state explosion and numerical
problems with solving very large systems of equations. On the other hand, the
increase of computer power and size of memory, as well as the development of
better software help us to overcome these problems.

This is why we are trying here to refine Markov models of router queueus. It
is well known that the distribution of the size of packets and self-similarity of the
input traffic have an impact on the transmission quality of service (determined by
transmission time, jitter, and loss probability); they influence also dynamics of
changes of number of packets waiting in routers to be forwarded. These issues are
usually investigated with the use of discrete-event simulations which in case of

© The Author(s) 2016
T. Czachórski et al. (Eds.): ISCIS 2016, CCIS 659, pp. 185–192, 2016.
DOI: 10.1007/978-3-319-47217-1_20

self-similar traffic demand very long runs and are time consuming, especially if we study transient states. Here we introduce to a Markov model details which were previously reserved to simulation models: a real distribution of IP packets and self-similar nature of packet flows. To obtain numerical results we use standard software: HyperStar [20] to approximate measured distributions with phase-type ones, enabling the use of Markov chains and Prism [11] to study transient states of a complex Markov model. We use also existing Markovian models of self-similar traffic [1]. With this purely engineering approach, we are able to construct more realistic than existing before models of IP queues and delays. The article is a continuation of [5] where we considered the queue length distributions at IP routers. Here we concentrate on the distribution of delays in these queues. The numerical study is based on more recent data.

2   Distribution of IP Packets

CAIDA, Center for Applied Internet Data Analysis [3], routinely collects traces on several backbone links. These monthly traces of one hour each are provided to interested researchers on request in pcap files containing payload-stripped, anonymised traffic. We used measurements of CAIDA coming from the link Equinix Chicago collected during one hour on 18 February 2016 having 22 644 654 packets belonging to 1 174 515 IPv4 flows, [4].

In a Markov model we should represent any real distribution with the use of a system of exponentially distributed phases (PH). Numerical PH fitting, e.g. with the use of Expectation Maximisation Algorithms, is a frequently investigated problem [2], and various tools exist, HyperStar [20], which we have chosen, is reported to be efficient at fitting spikes as in case of our distribution.

Figure 1 presents the cumulative distribution function (cdf) of IP packet lengths obtained from this trace and its approximation with the use of an hyper–Erlang distribution having three Erlang distributions with a variable number of phases, up to 3000. It demonstrates the quality of fitting as a function of the number of phases. To limit the number of states in the Markov model to follow, we have chosen the modest maximum number of phases to 300. The resulting Erlang distributions in parallel have 15, 4 and 300 phases, and its density function is (for $x \geq 0$)

$$f_B(x) = 0.05233 \frac{(0.01417)^{15} x^{15} e^{-0.01417x}}{14!} + 0.51162 \frac{(0.06067)^4 x^3 e^{-0.06067x}}{3!}$$
$$+ 0.43604 \frac{(0.20277)^{300} x^{299} e^{-0.20277x}}{299!}. \tag{1}$$

The largest approximation errors are at both extremities of the distribution, for small and large packets (e.g. the cdf is not equal 1 for the size of 1500 bytes). The mean of this distribution, i.e. mean packet size is 734.241 bytes. The same character has the distribution of service times, as the time to send a packet is proportional to its size, only phase parameters are rescaled. In numerical examples we assume that the buffer volume is equal to 64 mean packet size.

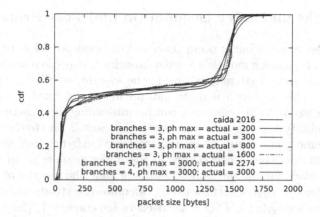

Fig. 1. The influence of the complexity of Markov model of a TCP packet size on the quality of the model

3 Self-similar Traffic

Since the mid 90s, with the collection of high-quality traffic measurements on several Ethernet LANs at the Bellcore Morristown Research and Engineering Center [12] and the statistical analysis of the collected data [13], self-similarity has become an important research domain [14]. In the following years the same statistical features have been confirmed by traffic measurements over different network and application scenarios. Moreover, various works highlighted the relevant impact of the long memory properties, typical of self-similar processes, on queueing dynamics; indeed, ignoring these phenomena leads to an underestimation of important performance measures, such as queue lengths at buffers and packet loss probability [8,10]. Therefore, it is necessary to take into account these features in realistic models of traffic.

Unfortunately, pure self-similar processes lack analytical tractability and only asymptotic results, typically derived in the framework of Large Deviation Theory, are available for simple queueing models (see, for instance, [15] and references therein). Therefore, many researchers investigated the suitability of Markovian models to describe traffic flows that exhibit self-similarity [6,21]. Different models have been proposed, but all works highlighted an important common conclusion: matching self-similarity is only required within the time scales of interest for the analysed system, e.g. [16].

As a result, more traditional and well investigated traffic models, such as Markov Modulated Poisson Processes (MMPPs), may still be used for modelling self-similar traffic. In this paper we focus on the model originally proposed in [1], and detailed in [6]. The model is simple: pseudo self-similar traffic can be generated as the superposition of a number of ON-OFF sources, a special case of two-state MMPPs, also known as Interrupted Poisson Processes, since the rate is zero when the modulating chain is in one of the two states (OFF state); we used five ON-OFF sources.

4   Remarks on Buffer Occupation and Loss Probability

Let us consider now the buffer occupation and the associated loss probability. In the majority of queueing models a system capacity is expressed as the maximum number of customers that may be inside the system, waiting in the queue or being served. This approach is quite natural in case of fixed-size packets (for instance, in case of ATM cells), but can be misleading in IP networks, due to the high variability of packets size, as described in Sect. 2, and to the fact that the amount of memory in a router is typically expressed in bytes. However, the queue length distribution when i packets are in the buffer permits us to determine if there is still place for the next one. Assuming that the lengths of the packets are independent, it is straightforward to calculate the steady-state conditional queue distribution $Q_i(x) = P(Q < x | i$ packets are enqueued) (for $i \geq 2$) as the i-fold convolution of the original distribution. Hence, we can easily calculate the probability that the queue length with i packets exceeds the volume V of the buffer and use this value as $p_{loss}(i)$, i.e. the probability that a packet is refused when there are already $i - 1$ packets in the buffer. The rate of the input flow is thus $\lambda(i) = \lambda(1 - p_{loss}(i))$.

It is worth mentioning that our approach introduces some kind of approximation: indeed, on one side we consider not the real length of the packets laying in the queue, but just the length distribution with which they have been generated. On the other side, the loss probability will depend also on the length of the arriving packet; so, if the queue is almost full for most of the time, it is likely it will mainly contain short packets and so the real queue length (in bytes) might be less, leading to an upper bound of the real p_{loss}. Instead, in case of lower utilisation, the queue length distribution seen by the arriving packet should be much closer to $Q_i(x)$ and hence our approximation works better.

5   Numerical Solutions, Transient States, Network Dynamics

Queueing models are usually limited to the analysis of steady states and popular Markovian solvers, as e.g. PEPS [17] are adapted to it. However, the intensities of real network traffic are perpetually changing; users send variable quantities of data, and traffic control algorithms interfere to avoid congestion (congestion window used in TCP is a good example).

Theoretically, for any continuous time Markov chain with transition matrix $\mathbf{Q}$ the Chapman-Kolmogorov equations

$$\frac{d\boldsymbol{\pi}(t)}{dt} = \boldsymbol{\pi}(t)\mathbf{Q}, \tag{2}$$

have the analytical transient solution $\boldsymbol{\pi}(t) = \boldsymbol{\pi}(0)e^{\mathbf{Q}t}$, where $\boldsymbol{\pi}(t)$ is the probability vector and $\boldsymbol{\pi}(0)$ is the initial condition. However, it is not easy to compute the expression $e^{\mathbf{Q}t}$ when $\mathbf{Q}$ is a large matrix. An efficient approach is to use a projection method, where the original problems is projected to a space (e.g. Krylov subspace) where it has considerably smaller dimension, solve it there and then re-transform this solution to the original state space [22]. It is implemented among others in a well known probabilistic model checker Prism [11]. We used Prism supplementing it with a preprocessor based on [18,19] to ease the formulation of more complex queueing models.

6 Response Time Distribution

Having the queue distribution $p(n)$, the response time (waiting time plus service time) probability density function (pdf) $f_R(x)$ is obtained as

$$f_R(x) = \sum_n p(n) f_B(x)^{*(n+1)}$$

where $f_B(x)$ is the pdf of service time distribution and $*(i)$ denotes i-fold convolution.

Figure 2 presents the comparison of response time distribution given by simulation and our model. Simulation was based on real traffic and packet size traces. In Markov model we used ON-OFF sources with the corresponding to measurements Hurst parameter (average of estimations made by several methods) and the described Hyper-Erlang distribution. In linear time scale the errors of the model are almost invisible. Therefore we use logarithmic time scale. The discrepancies are caused, amongst others, by the insufficient precision of approximation by the function in Eq. 1. It gives an under-representation of actual sizes of small packets, and a respective over-representation of large packets.

In numerical examples we use the validated above model to illustrate the impact of self-similarity, utilisation factor ϱ, and packet size distribution on the response time. In the examples the input flow starts at $t = 0$ and the queue is initially empty. Figures 3 and 4 present (i) the evolution of the mean response time as a function of time – time is normalized to the mean service time of

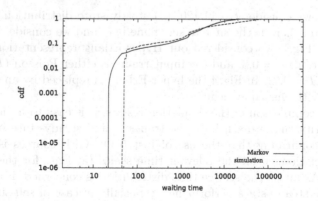

Fig. 2. Comparison of response time distribution given by simulation and Markov model, logarithmic time scale

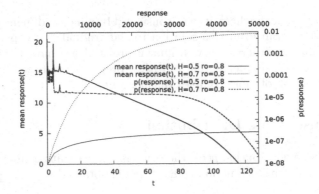

Fig. 3. Mean response time as a function of time and steady state distribution of response time for hyper-Erlang representation of service time distribution, $H = 0.5, 0.7$, $\varrho = 0.8$

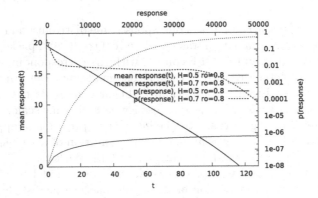

Fig. 4. Mean response time as a function of time and steady state distribution of response time for exponential service time distribution, $H = 0.5, 0.7$, $\varrho = 0.8$

a packet and we consider $t \in [0, 120]$ (ii) steady state distribution of response time – time unit here is the time to serve one byte and we consider the interval $[0, 50000]$. In Fig. 3 we considered our Hyper-Erlang representation of service time distribution, $\varrho = 0.8$, and the input traffic is either Poisson ($H = 0.5$) or self-similar ($H = 0.7$). In Fig. 4 the hyper-Erlang is replaced by an exponential distribution with the same mean.

From the comparison of the simulation results, it is easy to notice the effect of self-similarity that worsen both the transient and steady-state behaviour of the system, confirming that the use of just 5 ON-OFF sources is enough to capture correlation on all the relevant time scales (at least for the considered buffer size). As far as the service time distribution is concerned, it significantly influences the steady-state performance, especially in case of self-similar traffic (and hence for actual traffic flows). In other words, self-similarity and actual packet size distribution are relevant factors that must be taken into account in looking for realist traffic models.

7   Conclusions

In this work we proposed an approach that unifies in a Markovian model (i) a real IP packet distribution which is a basis to define both the losses due to a finite buffer volume and the service time distribution (ii) self similar traffic. The presented numerical examples, based on real traffic data collected by CAIDA a few months ago, confirm that our approach is feasible and may be used also to study transient behaviour of router delays.

Quantitative results may be obtained with the use of well known public software tools. As further work, we plan to apply our approach to Active Queue Management mechanisms.

References

1. Andersen, A.T., Nielsen, B.F.: A Markovian approach for modeling packet traffic with long-range dependence. IEEE J. Sel. Areas Commun. **16**(5), 719 (1998)
2. Buchholz, P., Kriege, J., Felko, I.: Input Modeling with Phase-Type Distributions, Markov Models: Theory and Applications. Springer, Heidelberg (2014)
3. http://www.caida.org/home/
4. https://data.caida.org/datasets/passive-2016/equinix-chicago/20160218-130000. UTC/
5. Czachórski, T., Domanski, A., Domanska, J., Rataj, A.: A study of IP router queues with the use of Markov models. In: Gaj, P., Kwiecien, A., Stera, P. (eds.) CN 2016. CCIS, vol. 608, pp. 294–305. Springer, Heidelberg (2016). doi:10.1007/978-3-319-39207-3_26
6. Domańska, J., Domański, A., Czachórski, T.: Modeling packet traffic with the use of superpositions of two-state MMPPs. In: Kwiecień, A., Gaj, P., Stera, P. (eds.) CN 2014. CCIS, vol. 431, pp. 24–36. Springer, Heidelberg (2014). doi:10.1007/978-3-319-07941-7_3
7. Fischer, W., Meier-Hellstern, K.: The Markov-modulated Poisson process (MMPP) cookbook. Perform. Eval. **18**(2), 149–171 (1993)
8. Gorrasi, A., Restino, R.: Experimental comparison of some scheduling disciplines fed by self-similar traffic. Proc. IEEE Int. Conf. Comm. **1**, 163–167 (2003)
9. Grossglauser, M., Bolot, J.C.: On the relevance of long-range dependence in network traffic. IEEE/ACM Trans. Netw. **7**(5), 629–640 (1999)

10. Kim, Y.G., Min, P.S.: On the prediction of average queueing delay with self-similar traffic. In: Proceedings of IEEE Globecom 2003, vol. 5, pp. 2987–2991 (2003)
11. Kwiatkowska, M., Norman, G., Parker, D.: PRISM 4.0: verification of probabilistic real-time systems. In: Gopalakrishnan, G., Qadeer, S. (eds.) CAV 2011. LNCS, vol. 6806, pp. 585–591. Springer, Heidelberg (2011). doi:10.1007/978-3-642-22110-1_47. www.prismmodelchecker.org/
12. Leland, W.E., Wilson, D.V.: High time-resolution measurement, analysis of LAN traffic: implications for LAN interconnection. In: INFOCOM 1991, Proceedings of the Tenth Annual Joint Conference of the IEEE Computer and Communications Societies. Networking in the 90s, vol. 3, pp. 1360–1366. IEEE (1991)
13. Leland, W.E., Taqqu, M.S., Willinger, W., Wilson, D.V.: On the self-similar nature of Ethernet traffic (extended version). IEEE/ACM Trans. Netw. 2(1), 1–15 (1994)
14. Loiseau, P., Gonçalves, P., Dewaele, G., Borgnat, P., Abry, P.V.-B., Primet, P.V.-B.: Investigating self-similarity and heavy-tailed distributions on a large-scale experimental facility. IEEE/ACM Trans. Netw. 18(4), 1261–1274 (2010)
15. Mandjes, M.: Large Deviations of Gaussian Queues. Wiley, Chichester (2007)
16. Nogueira, A., Salvador, P., Valadas, R., Pacheco, A.: Markovian modelling of internet traffic. In: Kouvatsos, D.D. (ed.) Next Generation Internet: Performance Evaluation and Applications. LNCS, vol. 5233, pp. 98–124. Springer, Heidelberg (2011). doi:10.1007/978-3-642-02742-0_5
17. PEPS. www-id.imag.fr/Logiciels/peps/userguide.html
18. Rataj, A., Wozna, B., Zbrzezny, A.: A translator of java programs to TADDs. Fundam. Inform. 93(1), 305 (2009)
19. Rataj, A.: More flexible models using a new version of the translator of java sources to times automatons J2TADD. Theor. Appl. Inform. 21(2), 107–114 (2009)
20. Reinecke, P., Krauß, T., Wolter, K.: HyperStar: phase-type fitting made easy. In: 9th International Conference on the Quantitative Evaluation of Systems (QEST) 2012, Tool Presentation, pp. 201–202, September 2012
21. Robert, S., Boudec, J.Y.L.: New models for pseudo self-similar traffic. Perform. Eval. 30(1-2), 57 (1997)
22. Stewart, W.: Introduction to the Numerical Solution of Markov Chains. Princeton University Press, Chichester (1994)

The Fluid Flow Approximation of the TCP Vegas and Reno Congestion Control Mechanism

Adam Domański[2], Joanna Domańska[1], Michele Pagano[3],
and Tadeusz Czachórski[1](✉)

[1] Institute of Theoretical and Applied Informatics, Polish Academy of Sciences,
Baltycka 5, 44–100 Gliwice, Poland
tadek@iitis.pl
[2] Institute of Informatics, Silesian Technical University,
Akademicka 16, 44–100 Gliwice, Poland
adamd@polsl.pl
[3] Department of Information Engineering, University of Pisa,
Via Caruso 16, 56122 Pisa, Italy
m.pagano@iet.unipi.it

Abstract. TCP congestion control algorithms have been design to improve Internet transmission performance and stability. In recent years the classic Tahoe/Reno/NewReno TCP congestion control, based on losses as congestion indicators, has been improved and many congestion control algorithms have been proposed. In this paper the performance of standard TCP NewReno algorithm is compared to the performance of TCP Vegas, which tries to *avoid* congestion by reducing the congestion window (CWND) size before packets are lost. The article uses fluid flow approximation to investigate the influence of the two above-mentioned TCP congestion control mechanisms on CWND evolution, packet loss probability, queue length and its variability. Obtained results show that TCP Vegas is a fair algorithm, however it has problems with the use of available bandwidth.

1 Introduction

In spite of the rise of new streaming applications and P2P protocols that try to avoid traffic shaping techniques and the definition of new transport protocols such as DCCP, TCP still carries the vast majority of traffic [10] and so its performance highly influences the general behavior of the Internet. Hence, a lot of research work has been done to improve TCP and, in particular, its congestion control features.

The first congestion control rules were proposed by Van Jacobson in the late 1980s [8] after that the Internet had the first of what became a series of congestion collapses (sudden factor-of-thousand drop in bandwidth). The first practical implementation of TCP congestion control is known as TCP Tahoe, while further evolutions are TCP Reno and TCP NewReno that better handles multiple losses in the same congestion window (CWND). The Reno/NewReno algorithm

© The Author(s) 2016
T. Czachórski et al. (Eds.): ISCIS 2016, CCIS 659, pp. 193–200, 2016.
DOI: 10.1007/978-3-319-47217-1_21

consists of the following mechanisms: Slow Start, Congestion Avoidance, Fast Retransmit and Fast Recovery. The first two, determining an exponential and linear grow respectively, are responsible for increasing CWND in absence of losses in order to make use of all the available bandwidth. Congestion is detected by packet losses, which can be identified through timeouts or duplicate acknowledgements (Fast Retransmit). Since the latter are associated to mild congestion, CWND is just halved (Fast Recovery) and not reduced to 1 packet as after a timeout. Hence, the core of classical TCP congestion control is the AIMD (Additive-Increase/Multiplicative-Decrease) paradigm. Note that this approach provides congestion control, but does not guarantee fairness [6].

The TCP Vegas was the first attempt of a completely different approach to bandwidth management and is based on congestion detection *before* packet losses [3]. In a nutshell (see Sect. 2 for more details), TCP Vegas compares the *expected rate* with the *actual rate* and uses the difference as an additional congestion indicator, updating CWND to keep the *actual rate* close to the *expected rate* and, at the same time, to be able of making use of newly available channel capacity. To this aim TCP Vegas introduces two thresholds (α and β), which trigger an Additive-Increase/Additive-Decrease paradigm in addition to standard AIMD TCP behavior. The article [12] shows TCP Vegas stability and congestion control ability, but, in competition with AIMD mechanism, it cannot fully use the available bandwidth.

The goal of our paper is to compare the performance of these two variants of TCP through fluid flow models. In more detail we investigated the influence of these two TCP variants on CWND changes and queue length evolution, hence also one-way delay and its variability (jitter). Moreover, we also evaluated the friendliness and fairness of the different TCP variants as well as their ability in using the available bandwidth in presence of both standard FIFO queues with tail drop and Active Queue Management (AQM) mechanisms in the routers.

Another important contribution of our work is that we considered also the presence of background traffic and asynchronous flows. In the literature, traffic composed of TCP and UDP streams has been already considered, but in most works (for instance, in [5,13]) all TCP sources had the same window dynamics and UDP streams were permanently associated with the TCP stream. Instead, in this paper, extending our previous work presented in [4], the TCP and UDP streams are treated as separate flows. Moreover, unlike [9] and [14], TCP connections start at different times with various values of initial CWND.

The rest of the paper is organized as follows. The fluid flow approximation models are presented in Sect. 2, while Sect. 3 discusses the comparison results. Finally, Sect. 4 ends the paper with some final remarks.

2   Fluid Flow Model of TCP NewReno and Vegas Algorithms

This section presents two fluid flow models of a TCP connection, based on [7,11] (TCP NewReno) and [2] (TCP Vegas). Both models use fluid flow approximation and stochastic differential equation analysis. The models ignore the TCP timeout mechanisms and allow to obtain the average value of key network variables.

In [11] a differential equation-based fluid model was presented to enable transient analysis of TCP Reno/AQM networks (flows throughput and queues length in bottleneck router). The authors also showed how to obtain ordinary differential equations by taking expectations of the stochastic differential equations and how to solve the resultant coupled ordinary differential equations numerically to get the mean behavior of the system. In more detail, the dynamics of the TCP window for the i-th stream are approximated by the following equation [7]:

$$\frac{dW_i(t)}{dt} = \frac{1}{R_i(t)} - \frac{W_i(t)W_i(t - R_i(t))}{2R_i(t - R_i(t))}p(t - R_i(t)) \qquad (1)$$

where:

- $W_i(t)$ – expected size of CWND (packets),
- $R_i(t) = \frac{q(t)}{C} + T_{p_i}$ – RTT (sec),
- $q(t)$ – queue length (packets),
- C – link capacity (packets/sec),
- T_{p_i} – propagation delay (sec),
- p – probability of packet drop.

The first term on the right hand side of the Eq. (1) represents the rate of increase of CWND due to incoming acknowledgments, while the second one models multiplicative decrease due to packet losses. Note that such model ignores the slow start phase as well as packet losses due to timeouts (a loss just halves the congestion window size) in accordance with a pure AIMD behavior, which is a good approximation of the real TCP behavior in case of low loss rates.

In solving Eq. (1) it is also necessary to take into account that the maximum values of q and W depend on the buffer capacity and the maximum window size (if the scale option is not used, 64 KB due to the limitation of the AdvertisedWindow field in TCP header). The dropping probability $p(t)$ depends on the discarding algorithm implemented in the routers (AQM vs. tail drop) and on the current queue size $q(t)$, which can be calculated through the following differential equation (valid for both models also in presence of background UDP traffic):

$$\frac{dq(t)}{dt} = \sum_{i=1}^{n_1} \frac{W_i(t)}{R_i(t)} + \sum_{i=1}^{n_2} U_i(t) - C\mathbf{1}_{q(t)>0} \qquad (2)$$

where $U_i(t)$ is the rate of the i-th UDP flow (with $U_i(t) = 0$ before the source starts sending packets), while n_1 and n_2 denote the number of TCP (NewReno or Vegas) and UDP streams, respectively. Note that the indicator function $\mathbf{1}_{q(t)>0}$ takes into account that packets are drawn at rate C only when the queue is not empty.

As already mentioned, classical TCP variants base their action on the detection of losses. The TCP Vegas mechanism, instead, tries to estimate the available bandwidth on the basis of changes in RTT and, every RTT, increases or decreases

CWND by 1 packet. To this aim, TCP Vegas calculates the minimum value of the RTT, denoted as R_{Base} in the following, assuming that it is achieved when only one packet is enqueued:

$$R_{Base} = \frac{1}{c} + T_p \tag{3}$$

Hence, the *expected rate*, which denotes the target transmission speed, is the ratio between CWND and the minimum RTT, i.e.:

$$Expected = \frac{W_i(t)}{R_{Base}} \tag{4}$$

while the *actual rate* depends on the current value $R(t)$ of the RTT:

$$Actual = \frac{W_i(t)}{R(t)} = \frac{W_i(t)}{\frac{q(t)}{c} + T_p} \tag{5}$$

The Vegas mechanism is based on three thresholds: α, β and γ, where α and β refer to the Additive-Increase/Additive-Decrease paradigm, while γ is related to the modified slow-start phase [3].

In more detail, for $Expected - Actual \leq \frac{\gamma}{R_{Base}}$ TCP Vegas is in the slow start phase, while for higher values of the difference we have the pure additive behavior: for $Expected - Actual \leq \frac{\alpha}{R_{Base}}$ the window increases by one packet for each RTT and for $Expected - Actual \geq \frac{\beta}{R_{Base}}$ the window decreases by the same amount. Finally, if $Expected - Actual$ is between the two thresholds α and β, CWND is not changed. Taking into account the definition of *expected* and *actual* rates given by Eqs. (4) and (5) respectively, it is possible to express the previous inequalities in terms of W_i, R and R_{Base}. Then, changes in the window are given by the formula:

$$\frac{dW_i(t)}{dt} = \frac{W(t - R(t)) * W(t - R(t))}{R(t)} p_0(t - R(t)) + \frac{1}{R(t - R(t))} p_1(t - R(t))$$
$$- \frac{1}{W(t - R(t))R(t - R(t))} p_2(t - R(t))$$

where

$$p_0 = \begin{cases} 1 & \text{for } \frac{W_i(R - R_{Base})}{R} \leq \gamma \\ 0 & \text{otherwise} \end{cases}, \qquad p_1 = \begin{cases} 1 & \text{for } \gamma \leq \frac{W_i(R - R_{Base})}{R} \leq \alpha \\ 0 & \text{otherwise} \end{cases}$$

and

$$p_2 = \begin{cases} 1 & \text{for } \frac{W_i(R - R_{Base})}{R} \geq \beta \\ 0 & \text{otherwise} \end{cases}.$$

3   Experimental Results

Our main goal is to show the behavior of the two completely different TCP mechanisms, taking into account various network scenarios in terms of amount of TCP flows as well as queue management disciplines (namely, standard FIFO with tail drop and RED, the best-known example of AQM mechanism). For numerical fluid flow computations we used a software written in Python and previously presented in [4]. During the tests we assumed the following TCP connection parameters:

- transmission capacity of bottleneck router: $C = 0.075$,
- propagation delay for i-th flow: $T_{p_i} = 2$,
- initial congestion window size for i-th flow (measured in packets): $W_i = 1, 2, 3, 4....$,
- starting time for i-th flow
- threshold values in TCP Vegas sources: $\gamma = 1$, $\alpha = 2$ and $\beta = 4$ (see [1,3]),
- RED parameters: $Min_{th} = 10$, $Max_{th} = 15$, buffer size = 20 (all measured in packets), $P_{max} = 0.1$, weight parameter $w = 0.007$,
- the number of packets sent by i-th flow (finite size connections).

Figures 1 and 2 present the CWND evolution and the buffer occupancy for different numbers of TCP Vegas connections. In more detail, Fig. 1(a) refers to a single TCP stream: after the initial slow start, the congestion avoidance phase goes on until the optimal window size is reached and then CWND is maintained at such level until the end of transmission. In case of two TCP connections (Fig. 1(b)), the evolution of CWND is identical for both streams and similar to the single source case (apart from a slightly lower value of the maximum CWND). The comparison between the two figures highlights the main disadvantage of TCP Vegas: the link underutilization. Indeed, under the same network conditions, the optimal CWND for one flow is only slightly less than the optimal CWND for each of the two flows.

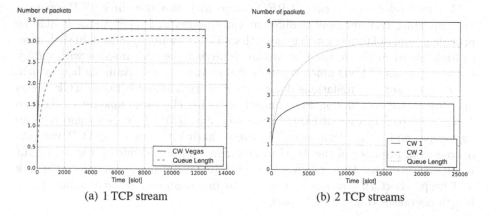

(a) 1 TCP stream   (b) 2 TCP streams

Fig. 1. TCP Vegas congestion window evolution — FIFO queue

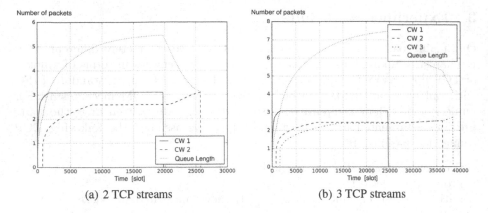

(a) 2 TCP streams　　　　　　　　(b) 3 TCP streams

Fig. 2. TCP Vegas congestion window evolution — RED queue

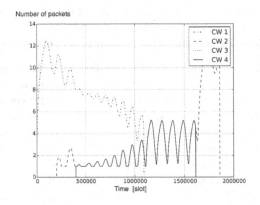

Fig. 3. TCP NewReno congestion window evolution — FIFO queue, 4 TCP streams

Figure 2 refers to the case of RED queue with two and three TCP streams. Streams start transmission at different time points and TCP Vegas is able to provide a level of fairness much greater than TCP NewReno. Indeed, in such case, as highlighted in Fig. 3, the first stream (starting the transmission with empty links) decreases CWND much slower and uses most of the available bandwidth.

The last set of simulations deals with the friendliness between TCP Vegas and NewReno, considering two connections with the same amount of data to be transmitted. In case of FIFO queue (see Fig. 4(a)), TCP NewReno is more aggressive and sends data faster. Uneven bandwidth usage by TCP variants decreases in presence of the AQM mechanism, as pointed out by Fig. 4(b). Our results confirm that the RED mechanism improves fairness in access to the link and keeps short the queues in routers (in our example, the maximum queue length decreases from 20 to 12 packets).

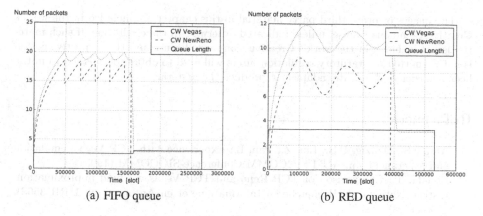

(a) FIFO queue   (b) RED queue

Fig. 4. TCP Vegas and NewReno congestion window evolution

4 Conclusions

The article evaluates by means of a fluid approximation the effectiveness of the congestion control of TCP NewReno and TCP Vegas.

The two TCP variants differ significantly in managing the available bandwidth. On one hand, TCP NewReno increases CWND to reach the maximum available bandwidth and eventually decreases it when congestion appears. This greedy approach clearly favors a stream which starts transmission when the link is empty. On the other hand, TCP Vegas increases CWND only up to a certain level to avoid the possibility of overloading. The disadvantage of this solution is the link underutilization: with a single stream TCP Vegas is conservative and may not use the total available bandwidth. However, in case of several competing streams, TCP Vegas mechanism shows its fairness: in presence of synchronous flows every stream uses the same share of the available bandwidth and even in case of streams starting transmission at different times a quite fair share of the network resources is still obtained.

Finally, the presented analysis permits to take into account finite-size flows and, unlike most works in this area, allows to start TCP transmission at any point of time with different values of the initial CWND (modern TCP implementation often starts with a window bigger than 1 packet). In other words, our approach makes possible the observation of TCP dynamics at such time when other sources start or end transmission.

References

1. Ahn, J.S., Danzig, P.B., Liu, Z., Yan, L.: Experience with TCP Vegas: emulation and experiment. In: ACM SIGCOMM Conference-SIGCOMM (1995)
2. Bonald, T.: Comparison of TCP Reno and TCP Vegas via fluid approximation. Institut National de Recherche en Informatique et en Automatique 1(RR 3563), 1–34 (1998)
3. Brakmo, L.S., Peterson, L.: TCP Vegas: end to end congestion avoidance on a global internet. IEEE J. Sel. Areas Commun. **13**(8), 1465–1480 (1995)
4. Domańska, J., Domański, A., Czachórski, T., Klamka, J.: Fluid flow approximation of time-limited TCP/UDP/XCP streams. Bull. Pol. Acad. Sci. Tech. Sci. **62**(2), 217–225 (2014)
5. Domański, A., Domańska, J., Czachórski, T.: Comparison of AQM control systems with the use of fluid flow approximation. In: Kwiecień, A., Gaj, P., Stera, P. (eds.) CN 2012. CCIS, vol. 291, pp. 82–90. Springer, Heidelberg (2012). doi:10.1007/978-3-642-31217-5_9
6. Grieco, L., Mascolo, S.: Performance evaluation and comparison of Westwood+, New Reno, and Vegas TCP congestion control. ACM SIGCOMM Comput. Commun. Rev. **34**(2), 25–38 (2004)
7. Hollot, C., Misra, V., Towsley, D.: A control theoretic analysis of RED. In: IEEE/INFOCOM 2001, pp. 1510–1519 (2001)
8. Jacobson, V.: Congestion avoidance and control. In: Proceedings of ACM SIGCOMM 1988, pp. 314–329 (1988)
9. Kiddle, C., Simmonds, R., Williamson, C., Unger, B.: Hybrid packet/fluid flow network simulation. In: 17th Workshop on Parallel and Distributed Simulation, pp. 143–152 (2003)
10. Lee, D., Carpenter, B.E., Brownlee, N.: Observations of UDP to TCP ratio and port numbers. In: Proceedings of the 2010 Fifth International Conference on Internet Monitoring and Protection, ICIMP 2010, pp. 99–104. IEEE Computer Society, Washington, DC (2010)
11. Misra, V., Gong, W., Towsley, D.: Fluid-based analysis of a network of AQM routers supporting TCP flows with an application to RED. In: Proceedings of ACM/SIGCOMM, pp. 151–160 (2000)
12. Mo, J., La, R., Anantharam, V., Walrand, J.: Analysis and comparison of TCP Reno and Vegas. In: Proceedings of IEEE INFOCOM, pp. 1556–1563 (1999)
13. Wang, L., Li, Z., Chen, Y.P., Xue, K.: Fluid-based stability analysis of mixed TCP and UDP traffic under RED. In: 10th IEEE International Conference on Engineering of Complex Computer Systems, pp. 341–348 (2005)
14. Yung, T.K., Martin, J., Takai, M., Bagrodia, R.: Integration of fluid-based analytical model with packet-level simulation for analysis of computer networks. In: Proceedings of SPIE, pp. 130–143 (2001)

Wireless Networks and Security

Baseline Analytical Model for Machine-Type Communications Over 3GPP RACH in LTE-Advanced Networks

Konstantin E. Samouylov[1], Yuliya V. Gaidamaka[1(✉)],
Irina A. Gudkova[1,2], Elvira R. Zaripova[1], and Sergey Ya. Shorgin[2]

[1] Applied Probability and Informatics Department, RUDN University,
6 Miklukho-Maklaya St., Moscow 117198, Russia
{ksam,ygaidamaka,igudkova,ezarip}@sci.pfu.edu.ru
[2] Federal Research Center "Computer Science and Control", Russian Academy
of Sciences, 44-2, Vavilova St., Moscow 119333, Russia
sshorgin@ipiran.ru

Abstract. Machine-type communication (MTC) is a new service defined by the 3rd Generation Partnership Project (3GPP) to provide machines to interact to each other over future wireless networks. One of the main problems in LTE-advanced networks is the distribution of a limited number of radio resources among enormously increasing number of MTC devices with different traffic characteristics. The radio resources allocation scheme for MTC traffic transmission in LTE networks is also standardized by 3GPP and implements the Random Access Channel (RACH) mechanism for transmitting data units from a plurality of MTC devices. Until now, there is a number of problems with the congestion in radio access network, as evidenced by a series of articles calling attention to the fact that more research is required, and even modification of the RACH mechanism in order to address drawbacks, exhibiting for example when a large number of devices are trying to access simultaneously. However, not many results have been obtained for the analysis, which allows to explore a variety of performance metrics of RACH mechanism on a qualitative level. In this paper the mathematical model in a form of the discrete Markov chain is built taking into account the features of the access procedure under congestion conditions and collisions. This baseline model allows to obtain the solution for key performance measures of RACH mechanism, such as the access success probability and the average access delay, in an analytical closed-form. Based on the proposed baseline model it is possible to obtain new results for the analysis of some modifications of RACH mechanism such as ACB (Access Class Baring).

Keywords: LTE-advanced · Machine-type communications · Random access channel · Markov chain · Access success probability · Average access delay

The reported study was funded by RFBR and Moscow city Government according to the research project No. 15-37-70016 mol_a_mos, by RFBR according to the research projects No. 14-07-00090, 15-07-03051, and by Ministry of Education and Science of the Russian Federation (President's Scholarship No. 2987.2016.5).

T. Czachórski et al. (Eds.): ISCIS 2016, CCIS 659, pp. 203–213, 2016.
DOI: 10.1007/978-3-319-47217-1_22

1 Introduction

In recent years, a huge number of technological devices appeared in the market that support various applications associated with data transfer automatically. In this perspective, a key role will be played by machine-type communications (MTC), which is a new concept where devices exchange data without any (or minimal) human intervention [1]. MTC is expected to open up unprecedented opportunities for telecom operators in the various fields of the new digital economy (home and office security and automation, smart metering and utilities, maintenance, building automation, automotive, healthcare and consumer electronics, etc.), and, therefore, will be one of the economic foundations of emerging 5G wireless networks [2, 3]. As in the case of any new technology, the analysis of the impact of MTC traffic features requires modification of both classical and modern methods [4–6].

Conventional wireless communication technologies, including 3GPP LTE network, do not allow establishing effectively machine-to-machine (M2 M) connections between a large number of interacting MTC devices. One possible solution of the problem is based on the use of random access (RA) procedure [7, 8]. The advantage of this method is that the MTC devices can access to the radio access channel (RACH), regardless of their arrangement and centralized management.

It is well known that an overload on the RACH level can lead to overload in the entire LTE network. Feature of the M2 M traffic that differs substantially from the traditional H2H traffic is that existing mechanisms cannot effectively overcome RA procedure overload. MTC devices such as fire detectors usually send small amounts of data periodically while operating in the normal mode. However, in the case of emergency MTC devices generate burst traffic, which can cause overloading [9, 10]. In the case of high network traffic access delay increases significantly, and this can be critical in various emergency situations [7]. Some other features of M2 M traffic transmission were considered in [10–19] taking into account problems of optimal radio resources allocation [11–15], overload control mechanisms based on Access Class Barring (ACB) schemes [10, 14] and other congestion control problems [16, 17].

The purpose of this paper is the analytical modeling of the access procedure which is able to support the simultaneous access of MTC devices. According to [7] the reference scheme of the procedure consists of a four-message handshake between the accessing devices and the base station. In the same 3GPP technical report main measures to RACH capacity evaluation for MTC are specified: collision probability, access success probability, access delay, the number of preamble transmissions to perform a random access procedure successfully, the number of simultaneous preamble transmissions in an access slot.

There are many papers devoted to modeling and simulation of RACH procedure, e.g. interesting results are obtained in [2], which also provides a review of known works on this issue. However, not many analytical models are known, which allow exploring main RACH performance metrics [7] on a qualitative level. We highlight [18], where the formulas for the calculation of these metrics were obtained. Unlike to known results, the objective of this study is to obtain a closed-form solution, which depends on the minimum number of RACH procedure parameters and is easy for calculation. This paper is an extension of [19], where the approach to analytical modeling using Markov chain apparatus was proposed and the Monte Carlo simulation model was developed. In contrast to [19], this paper concentrates on the analytical model of a random access procedure in LTE cell and focuses on two metrics for RACH capacity evaluation – the access success probability and the average access delay in the presence of collisions and physical channel congestion.

The rest of the paper is organized as follows. In Sect. 2 we shortly describe RACH signaling reference scheme, simultaneously discuss notations of the mathematical model and introduce its core assumptions. In Sect. 3, formulas for calculating key metrics in closed form are obtained. Further, in Sect. 4 main performance measures calculating is illustrated via the numerical example. Finally, we conclude the paper in Sect. 5.

2 Random Access Procedure, Model Notation and Assumptions

In this section we consider RACH procedure that is the initial synchronization process between user equipment (UE) and the base station eNB while data exchange performs over Physical RACH (PRACH) in LTE network [7]. Since UEs' attempt for data transmission can be performed randomly and the value of distance to the eNB is unknown, requests for synchronization from various UEs should come with different delays, which is estimated by the level of incoming PRACH signal by eNB.

Widely known RACH procedure defines the sequence of signaling messages transmitted between the UE and the eNB. The procedure begins with a random access preamble transmission to the eNB (Msg 1) by means of one of available PRACH slots (RACH opportunity). The information about slots is broadcasted by the eNB in System Information Block messages. The number of RACH opportunities and the number of preambles depend on the particular LTE network configuration.

After preamble sending the UE waits for a random access response (RAR) (Msg 2) from the eNB within the time interval called a response window. RAR message transmitting over Physical Downlink Control Channel (PDCCH) contains a resource grant for transmission of the subsequent signaling messages. If after the response window is over the UE has not received Msg 2, it means that a collision occurs. The collision of a preamble transmission may occur when two or more UEs select the same preamble and send it at same time slot. In the case of a collision the UE should repeat preamble transmission attempt after a response window. If a preamble collision occurs, the eNB will not send RAR message to all UEs, which have chosen the same preamble. In that case, preambles will be resent after the time interval called the backoff window. If series of collisions occur for a UE after the number of failed attempts exceeds the preamble attempts limit, the RACH procedure is recognized failed.

In the case of successful preamble transmission after receiving Msg 2 from the eNB and RAR processing time, the UE sends connection request (Msg 3) to the eNB using resources of Physical Uplink Shared Channel (PUSCH) [20]. RACH procedure is considered completed after the UE received a contention resolution message (Msg 4) from the eNB. Hybrid automatic repeat request (HARQ) procedure guarantees a successful transmission of Msg 3/Msg 4. HARQ procedure provides a limit in Msg 3/Msg 4 sequential transmission attempts. If the limit is reached UE should start a new RACH procedure by sending a preamble.

Making a number of simplifying assumptions for the RACH procedure, we introduce below the basic notation and build a mathematical model in the form of a discrete Markov chain according to [19]. The time interval between the first RA attempt and the completion of the random access procedure is called an access delay [7]. To analyze this parameter we propose a mathematical model in the form of discrete Markov chain that follows the steps of RACH procedure. The state of the Markov chain determines the number of preamble attempt collisions and the number of sequential Msg 3/Msg 4 transmission attempts. With this model the access delay for each state of the Markov chain can be calculated by summing up the corresponding time intervals introduced below:

Δ_1^1 – waiting time for a RACH opportunity to transmit a preamble;

Δ_1^2 – preamble transmission time;

Δ_1^3 – preamble processing time at the eNB;

Δ_1^4 – RAR response window;

$\Delta_1 := \Delta_1^1 + \Delta_1^2 + \Delta_1^3 + \Delta_1^4$ – time from the beginning of RACH procedure until sending message Msg 3 or resending a preamble;

Δ_2 – backoff window;

Δ_3 – RAR processing time;

Δ_4 – time for Msg 3 transmission, waiting for Msg 4, and Msg 4 processing.

The model notation is illustrated in message sequence diagram for access success (Fig. 1) and access failure (Fig. 2).

In the case of reliable connections the access delay is equal to the sum of the mentioned above variables Δ_i, $i = 1, 3, 4$. If a collision occurs or connection is

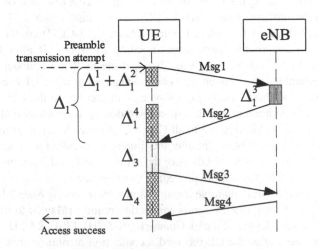

Fig. 1. Message sequence diagram for successful access

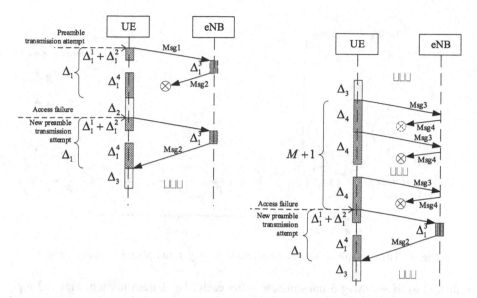

Fig. 2. Message sequence diagram for access failure due to (a) preamble collision (b) contention resolution message retransmission

unreliable the number of retransmissions is limited by $N = 9$ for Msg 1 and by $M = 4$ for Msg 3 [7]. Let p denote the collision probability, defined as the ratio between the number of occurrences when two or more MTC devices send a random access attempt using exactly the same preamble and the overall number of opportunities (with or without access attempts) in the period [7]. This value depends on the number of MTC devices at eNB coverage area, on intensity γ of incoming calls and on LTE network configuration. Also, let g denote the HARQ retransmission probability for Msg 3/Msg 4, and thus we entered all the variables needed further for obtaining formulas for calculation of the access success probability and the average access delay.

3   The Model and Results in a Closed Form

The formalization of the above-described RA procedure according to [19] is given by the absorbing discrete-time Markov' chain $\{\xi_i, i = 0, \ldots, (N+1)(M+1)+1\}$ with the finite state space

$$\mathbf{X} = \{(n, m, k), n = 0, \ldots, N, m = 0, \ldots, M, k = 0, \ldots, n\} \cup \{\omega, \upsilon\},$$

initial state $(0, 0, 0)$, and two absorbing states ω and υ. The initial state $(0, 0, 0)$ represents the beginning of the procedure followed by the first RA attempt, the absorbing state ω stands for the access success, and the absorbing state υ stands for the access failure. Other states denoted by (n, m, k), where n is the number of Msg 1 (preamble) retransmissions, m is the number of Msg 3 retransmissions after the last successful Msg 1 transmission, and k stands for the number of successful Msg 1 transmissions

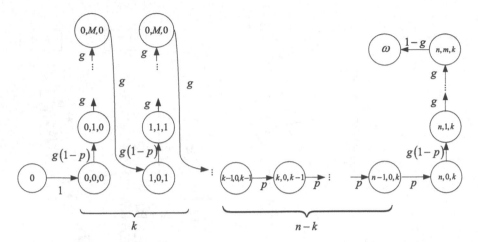

Fig. 3. The example of successful access with Msg 1 and Msg 3 retransmissions

followed by $M + 1$ Msg 3 transmissions after each Msg 1 transmission. Figure 3 represents one of possible paths from state $(0, 0, 0)$ to state (n, m, k) for successful access.

Note, that the access delay for RA procedure is defined as the time interval from the instant when a UE sends its first random access preamble until the UE receives the random access response [7]. In the paper, we focus on the average value D of the access delay. To calculate it we consider all possible scenarios of the RA procedure, i.e. different number of Msg 1 and Msg 3 retransmissions for different combinations of messages' sequences that influence on the overall access delay. For example, in the case of the successful access without any collision the sequence is Msg1 $\rightarrow$ Msg2 $\rightarrow$ Msg3 $\rightarrow$ Msg4. For the successful access with two retransmissions of message Msg1 and without Msg3 retransmissions the sequence looks like Msg1 $\rightarrow$ Msg1 $\rightarrow$ Msg1 $\rightarrow$ Msg2 $\rightarrow$ Msg3 $\rightarrow$ Msg4.

Note that we do not distinguish between two paths having the same delay between the first RA attempt and the same intermediate state (n, m, k), if the paths differ only Msg 1/Msg 3 positions. For example, the following message sequences (Msg 2 and Msg 4 are omitted) have the equal delays:

$$Msg1 \rightarrow Msg1 \rightarrow Msg3 \rightarrow \ldots \rightarrow Msg3 \rightarrow Msg1 \rightarrow Msg3$$

and

$$Msg1 \rightarrow Msg3 \rightarrow \ldots \rightarrow Msg3 \rightarrow Msg1 \rightarrow Msg1 \rightarrow Msg3.$$

Under these assumptions, the probability $P(n, m, k)$ of Markov chain $\{\xi_i\}$ visiting state (n, m, k) when starting from state $(0, 0, 0)$ is determined by the formula

$$P(n, m, k) = p^{n-k} C_n^k \left((1 - p)g^{M+1}\right)^k (1 - p)g^m, \quad (n, m, k) \in \mathbf{X}. \tag{1}$$

The first multiplier p^{n-k} stands for $n-k$ Msg 1 collisions, the multiplier $((1-p)g^{M+1})^k$ stands for k successful Msg 1 transmissions each followed by $M+1$ Msg 3 transmissions, the multiplier $(1-p)g^m$ stands for a unique successful Msg 1 transmission followed by m Msg 3 retransmissions, and the binomial coefficient C_n^k reflects the number of k combinations (successful Msg 1 transmissions) of an n set (Msg 1 retransmissions).

The probabilities of being absorbed in the states ω and υ when starting from state $(0,0,0)$ are

$$P(\omega) = \sum_{(n,m,k)\in\mathbf{X}} P(n,m,k)\cdot(1-g) = 1-\left(p+(1-p)g^{M+1}\right)^{N+1}, \tag{2}$$

$$P(\upsilon) = 1-P(\omega) = \left(p+(1-p)g^{M+1}\right)^{N+1}. \tag{3}$$

Note, that these probabilities for the RA procedure stand for the access success probability $P(\omega)$ and for the access failure probability $P(\upsilon)$.

For successful random access procedure we denote $Q(n,m,k)$ the probability that the RA procedure will be completed right after state (n,m,k), i.e. there will not be any further Msg1/Msg3 collisions. Let $D(n,m,k)$ be the corresponding access delay under the condition that random access procedure is successful.

The access delay $D(n,m,k)$ can be calculated as follows

$$\begin{aligned}D(n,m,k) &= (n-k)(\Delta_1+\Delta_2)+k(\Delta_1+\Delta_3+M\Delta_4)+\Delta_1+\Delta_3+(m+1)\Delta_4 \\ &= (\Delta_1+\Delta_2)\cdot n+\Delta_4\cdot m+(\Delta_3+M\Delta_4-\Delta_2)\cdot k+\Delta_1+\Delta_3+\Delta_4.\end{aligned} \tag{4}$$

Form the definition of probability $Q(n,m,k)$ we get the formula

$$\begin{aligned}Q(n,m,k) &= \text{P\{no Msg1/Msg3 collisions after state }(n,m,k)\mid\text{ successful access\}} \\ &= \frac{\text{P\{no Msg1/Msg3 collisions after state }(n,m,k),\ \text{ successful access\}}}{\text{P\{successful access\}}} \\ &= \frac{\text{P\{no Msg1/Msg3 collisions after state }(n,m,k)\}}{\text{P\{successful access\}}} = \frac{P(n,m,k)\cdot(1-g)}{P(\omega)}.\end{aligned} \tag{5}$$

Now, taking into account that the average RA delay, which is calculated only for successfully accessed MTC devices, is determined by the formula

$$D = \sum_{(n,m,k)\in\mathcal{X}} Q(n,m,k)D(n,m,k), \tag{6}$$

and taking into account (1)–(5), we finally obtain the formula to calculate the average access delay in closed form

$$D = (\Delta_1 + \Delta_2) \cdot \frac{C}{(1-p)(1-g^{M+1})} \left(1 - (N+1)C^N + NC^{N+1}\right)$$
$$+ \Delta_4 \cdot \frac{1 - (M+1)g^M + Mg^{M+1}}{1-g} \frac{g(1-C^{N+1})}{1-g^{M+1}} \tag{7}$$
$$+ (\Delta_3 + M\Delta_4 - \Delta_2) \cdot \frac{g^{M+1}}{1-g^{M+1}} \left(1 - (N+1)C^N + NC^{N+1}\right)$$
$$+ (\Delta_1 + \Delta_3 + \Delta_4) \cdot \left(1 - C^{N+1}\right),$$

where $C = p + g^{M+1}(1-p)$.

The numerical example in the next section illustrates the application of the formulas obtained for calculation the access success probability and the average access delay with given collision probability.

4 Numerical Example

We present an example of analysis of a single LTE FDD cell on 5 MHz supporting M2M communications to illustrate some performance measures for RACH with initial data closed to real ones [7, 9, 10, 18, 19].

In LTE, the RACH could be configured to occur once every subframe up to once every other radio frame. As in [7] we assume that the PRACH configuration index is equal to 6, and then for FDD cell we have 1st and 6th subframes of every frame for RACH opportunity, so the RACH occurs every 5 ms, that gives us 200 RACH opportunities per second. The total number of RACH preambles available in LTE is 64. A number of them are normally reserved for contention free RA procedure (i.e. for intra-system handover or downlink data arrival with lost synchronization), the rest are used for contention based RA procedure. According to [7] we assume that 10 preambles are configured to be dedicated for handovers, therefore, the other 54 can be used contention based random access.

For the scenario with a large number of UEs with RA procedure in the cell and uniformly distributed arrival of RACH requests the collision probability is given by [9]

$$p = 1 - e^{-\gamma/(54*200)}. \tag{8}$$

Maximum number of preamble transmission is equal to 10, hence $N = 9$. Maximum number of Msg 3 retransmissions $M = 4$ [7]. The terms of the sum in (7) are given below: $\Delta_1^1 = 2{,}5$ ms; $\Delta_1^2 = 1$ ms; $\Delta_1^3 = 2$ ms; $\Delta_1^4 = 5$ ms; $\Delta_2 = 20$ ms; $\Delta_3 = 5$ ms; $\Delta_4 = 6$ ms. The calculation were done for 4 values of the HARQ retransmission probability for Msg 3/Msg 4 $g = 0.02$; 0.5; 0.8; 0.95.

Typically, e.g. [7, 18], RACH performance metrics are analyzed vs the number of MTC devices per cell with maximum of 30 000. In the numerical example we analyze target metrics vs the collision probability p, receiving its value from the formula (8) with given random access intensity γ. Namely the value of γ indicates the number of MTC devices in the cell, but it does not reflect the number explicitly. For example,

$\gamma = 25\ 000$ attempts per second corresponds to the case of overload with the collision probability p about 0.9. By changing the collision probability p from 0 to 1 we compute the access success probability $P(\omega)$ using (2) and the average access delay D using (7).

Figure 4 introduces plots illustrating the access success probability $P(\omega)$ for four values of the HARQ retransmission probability g. The plots show that with g less than 0.5 even for $\gamma = 10\ 000$ attempts per second ($p = 0.6$) the access success probability is close to 1.

Figure 5 indicates that the average access delay D varies significantly with the changing of the collision probability p and even for minor g can reach values exceeding 160 ms due to a significant number of preamble retransmissions.

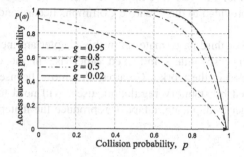

Fig. 4. Access success probability

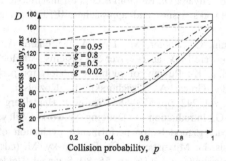

Fig. 5. Average access delay

5 Conclusion

In this paper we addressed a RACH procedure for service M2 M traffic in LTE cell and introduced a mathematical model in the form of discrete Markov chain. Note that the access success probability is critical for applications such as fleet management service, when a large number of taxis equipped with fleet management devices gather in a cell, for example near the airport. Another measure, the average access delay, is critical for earthquake monitoring applications, because even tens of milliseconds are very important for an earthquake alarm. The proposed model allows calculating both mentioned performance measures for LTE FDD and TDD cell, UMTS FDD or UMTS 1.28Mcps TDD.

An interesting task for future study is to derive a formula for the cumulative distribution function (CDF) of the access delay between the first RA attempt and the completion of the random access procedure, for the successfully accessed MTC devices. Another important problem is the construction of analytical models of the overload control mechanisms based on Access Class Barring (ACB) schemes.

References

1. Hasan, M., Hossain, E., Niyato, D.: Random access for machine-to-machine communication in LTE-advanced networks: issues and approaches. IEEE Commun. Mag. **51**(6), 86–93 (2013)
2. Condoluci, M., Araniti, G., Dohler, M., Iera, A., Molinaro, A.: Virtual code resource allocation for energy-aware MTC access over 5G systems. Ad Hoc Netw. **43**, 3–15 (2016)
3. Gorawski, M., Grochla, K.: Review of mobility models for performance evaluation of wireless networks. In: Gruca, A., Czachórski, T., Kozielski, S. (eds.) Man-Machine Interactions 3. AISC, vol. 242, pp. 573–584. Springer, Heidelberg (2014)
4. Gelenbe, E., Pujolle, G.: Introduction to Queueing Networks. Wiley, New York City (2000)
5. Czachórski, T.: Queueing models for performance evaluation of computer networks - transient state analysis. In: Mityushev, V.V., Ruzhansky, M. (eds.) Analytic Methods in Interdisciplinary Applications, vol. 116, pp. 55–80. Springer, Heidelberg (2015). PROMS
6. Andreev, S., Hosek, J., Olsson, T., Johnsson, K., Pyattaev, A., Ometov, A., Olshannikova, E., Gerasimenko, M., Masek, P., Koucheryavy, Y., Mikkonen, T.: A unifying perspective on proximity-based cellular-assisted mobile social networking. IEEE Commun. Mag. **54**(4), 108–116 (2016)
7. GPP TR 37.868 – Study on RAN Improvements for Machine-type Communications. Release 11. September 2011 (2011)
8. GPP LTE Release 10 & beyond (LTE-Advanced)
9. GPP R1-061369: LTE Random-access Capacity and Collision Probability, Ericsson, RAN1#45, May 2006 (2006)
10. Beale, M.: Future challenges in efficiently supporting M2 M in the LTE standards. In: Proceedings of the 10th Wireless Communications and Networking Conference WCNCW 2012, Paris, France, pp. 186–190. IEEE (2012)

11. Hossain, M., Niyato, D., Han, Z.: Dynamic Spectrum Access and Management in Cognitive Radio Networks. Cambridge University Press, Cambridge (2009)
12. Borodakiy, V.Y., Buturlin, I.A., Gudkova, I.A., Samouylov, K.E.: Modelling and analysing a dynamic resource allocation scheme for M2 M traffic in LTE networks. In: Balandin, S., Andreev, S., Koucheryavy, Y. (eds.) NEW2AN 2013 and ruSMART 2013. LNCS, vol. 8121, pp. 420–426. Springer, Heidelberg (2013)
13. Buturlin, I.A., Gaidamaka, Y.V., Samuylov, A.K.: Utility function maximization problems for two cross-layer optimization algorithms in OFDM wireless networks. In: Proceedings of the 4th International Congress on Ultra Modern Telecommunications and Control Systems ICUMT-2012, pp. 63–65. IEEE (2012)
14. Gudkova, I., Samouylov, K., Buturlin, I., Borodakiy, V., Gerasimenko, M., Galinina, O., Andreev, S.: Analyzing impacts of coexistence between M2 M and H2H communication on 3GPP LTE system. In: Mellouk, A., Fowler, S., Hoceini, S., Daachi, B. (eds.) WWIC 2014. LNCS, vol. 8458, pp. 162–174. Springer, Heidelberg (2014)
15. Shorgin, S., Samouylov, K., Gaidamaka, Y., Chukarin, A., Buturlin, I., Begishev, V.: Modeling radio resource allocation scheme with fixed transmission zones for multiservice M2 M communications in wireless IoT infrastructure. In: Nguyen, N.T., Trawiński, B., Kosala, R. (eds.) ACIIDS 2015. LNCS, vol. 9012, pp. 473–483. Springer, Heidelberg (2015)
16. Cheng, M., Lin, G., Wei, H.: Overload control for machine-type-communications in LTE-advanced system. IEEE Commun. Mag. 50(6), 38–45 (2012)
17. Dementev, O., Galinina, O., Gerasimenko, M., Tirronen, T., Torsner, J., Andreev, S., Koucheryavy, Y.: Analyzing the overload of 3GPP LTE system by diverse classes of connected-mode MTC devices. In: Proceedings of the IEEE World Forum on Internet of Things 2014, pp. 309–312 (2014)
18. Wei, C.-H., Bianchi, G., Cheng, R.-G.: Modelling and analysis of random access channels with bursty arrivals in OFDMA wireless networks. IEEE Trans. Wireless Commun. 14(4), 1940–1953 (2015)
19. Borodakiy, V.Y., Samouylov, K.E., Gaidamaka, Y.V., Abaev, P.O., Buturlin, I.A., Etezov, S.A.: Modelling a random access channel with collisions for M2 M traffic in LTE networks. In: Balandin, S., Andreev, S., Koucheryavy, Y. (eds.) NEW2AN/ruSMART 2014. LNCS, vol. 8638, pp. 301–310. Springer, Heidelberg (2014)
20. GPP TS 36.211 - Evolved Universal Terrestrial Radio Access (E-UTRA) - Physical Channels and Modulation (ver. 13.1.0 Release 13 April 2016) (2016)

Global Queue Pruning Method for Efficient Broadcast in Multihop Wireless Networks

Sławomir Nowak, Mateusz Nowak, Krzysztof Grochla$^{(\boxtimes)}$, and Piotr Pecka

Institute of Theoretical and Applied Informatics,
Polish Academy of Sciences, Gliwice, Poland
{emanuel,mateusz,kgrochla,piotr}@iitis.pl

Abstract. The article proposes a novel broadcast algorithm for multihop wireless networks. We compare three reference algorithms: Counter Based, Scalable Broadcast and Dominant Pruning, and propose a novel Global Queue Pruning method, which limits the overhead of the transmission and provides assurance of the delivery of the messages to every node in the network. The developed algorithm creates the logical topology that consists of lower number of forwarders in comparison to the previous methods, the paths are shorter, and the 100 % coverage is guaranteed. This is achieved with the higher cost of propagation of the topology information in the initialisation phase.

Keywords: Mesh networks · Multihop broadcast · Broadcast storms · Dominant pruning

1 Introduction

Smart devices, which communicate with each other and are part of the Internet of Things or IoT, become more and more popular. Advanced Metering Infrastructure (AMI) is a popular application of IoT devices, deployed to monitor the energy or water use. The IoT devices passing data from physical objects to the digital world are more and more widely used. The IoT networks consist of thousands of devices, creating a complex, multihop network. This causes increasingly stronger need to develop methods for the management of large networks of relatively simple devices, and need of development of reliable communication method for them. It is important to propose effective methods for broadcast and multicast communication, as sending messages, directed to all nodes or big groups of nodes is a popular case in AMI and IoT networks.

IoT networks differ in theirs specifics. Depending on their purpose, their topology may be static or dynamic. The number and location of nodes also may vary, which results in different characteristics of connection graph – dense or sparse, uniform or clustered. The source of power (battery or power line) is also the factor influencing chosen methods of communication. Most of the multicast and broadcast transmissions is directed from the designated central point to all nodes and from single node (unicast) to the central point.

© The Author(s) 2016
T. Czachórski et al. (Eds.): ISCIS 2016, CCIS 659, pp. 214–224, 2016.
DOI: 10.1007/978-3-319-47217-1_23

The multicast or broadcast transmission in multihop wireless networks requires the selection which nodes shall forward messages and act as intermediate point of communication, forwarding packets coming from other nodes (referenced as forwarders in further part of the paper). The remaining nodes are only receiving messages and act as the communication endpoint. The selection which nodes should forward the data and which should only receive it is a challenging task. A few algorithms have been proposed in the literature, however previous papers refer to a simple topologies with small average number of neighboring nodes (2–5). In wireless AMI networks the average number of nodes to which a node can communicate is considerably higher [8].

The simplest solution for broadcast transmission is flooding, the concept in which every incoming packet is sent through every outgoing link except the one it arrived on [9]. Flooding utilizes every path through the network, so it guarantees 100 % cover (if link transmissions are 100 % reliable) and it will also use the shortest path. This algorithm is also very simple to implement but has disqualifying disadvantages: can be costly in terms of wasted bandwidth and can impose a large number of redundant transmissions. Flooding is also not practical in dense networks, as it greatly increases the required transmission time [4].

Another method is to select the Connected Dominant Set (CDS) of nodes (forwarding nodes, forwarders). It was proved [2] that the optimal selection of CDS is a NP hard problem even if the whole network topology is known. The forwarder can be selected dynamically or statically [5]. In the static approach a global algorithm determines the status (forwarder/non forwarder) of each node and the level is set. In the dynamic approach the status is decided "on-the-fly" based on local node information, and the state can be different for every transmitted message. In the [5] interesting algorithm was presented using static approach and local topology information, however the node position information is assumed.

In this work we concentrate on an AMI network use case, with meters communicating by wireless interfaces. Meters are located within the buildings and they have a power supply. Changes in placement of sensor nodes are rare and done under control of network operator, so there is no need of automatic reconfiguration of network topology. There are no limitations of battery power, but it is the necessity of reliable communication and possibly optimal usage of network resources (bandwidth). We assume that a designated control node is distinguished, which typically has the access to a backhaul interface and forwards the traffic to and from the Internet to the AMI network.

We propose a novel algorithm (Global queue pruning) for forwarding nodes selection, which outperforms the solutions available in the literature. The proposed algorithm is compared to the three representative methods of forwarding nodes selection and evaluated through an extensive simulation study. Previous studies on multicast algorithm pointed also the disadvantages of the popular broadcast solution for RPL protocol (IP level multicast). The main problem is, that RPL it not designed to fit the specific of our network (root to sensor traffic) [10]. The RPL broadcast results in many overlapping transmission (particularly problematic for dense urban area where the level of overlap is high). To address the needs of our network we decided to control the message forwarding on the application level to replace the RPL build in the 6lowPAN protocol and their multicast mechanism.

2   The Problem Formulation

The layer 2 protocols determine the connectivity between nodes in the wireless network. This defines the topology of a network. In wireless sensor networks to send a message between two distant nodes it is usually necessary to use the intermediate nodes. The path of communication is composed of a sequence of such intermediate, forwarding nodes, called forwarders. Forwarders receive messages and under conditions of given algorithm, can retransmit it. Every forwarder in the network can be described by a level. The level is the number of hops from a central point to a node in the range of the forwarder. The forwarder level 0 is central control point – original source of broadcast messages or final receiver of messages from the nodes (Fig. 1).

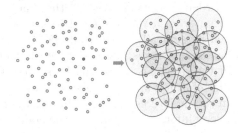

Fig. 1. Process of creating logical topology and selecting forwarder's nodes

Connected forwarders, from lower levels to higher create the logical topology of the network (spanning tree called Connected Dominant Set [5]). This set can be used both to the unicast, selective multicast and broadcast communication from control node to all nodes in the network. It is possible to distinguish more than one path, according to different selection of forwarders it is possible to compare the resulting logical topologies. The simplest metric to compare different topologies created in given network, used in this article, is the highest level of the forwarder in the path, what is equal to the maximum number of hops in the network. The average forwarder level is proportional to the average time of message propagation. We assume that the topology is determined in an initialization phase, in which the forwarders are selected which precedes the actual communication phase.

The goal of this work is to define method for "near to optimal" selection of the forwarders. A forwarder may only forward a packet once (to avoid infinite loops) and all nodes shall receive the packet in no-failure conditions (the forwarders don't fail and the topology of connections don't change during the transmission). The assumption is to achieve minimum broadcast overhead, respecting the possible nodes and link failures and to support the selective broadcast and multicast.

3 Reference Solutions

There are several solutions that can be used to route messages in the network and to select the forwarders. Besides the mentioned above flooding algorithms, there is a number of more complex and efficient methods. Some methods are based on the location knowledge (e.g. position for GPS signal), but those methods are not subject of analysis, as it was not assumed that the location information is detailed enough to be used and is available. Another group of methods is based on the neighbour knowledge methods. Knowing the neighbourhood of a node can be used to select a forwarders. Two approaches are possible: local (only local, or 2 hops neighbourhood is known) and global (the global information about nodes neighbourhood is known). Using the probabilistic methods it is possible to distinguish a set of forwarders, in which the randomization is used to decide on the packer retransmission (forward). We decided to implement the three reference solutions: one example of probabilistic method: counter based (CB) [3] and two neighbour knowledge methods: scalable broadcast algorithm (SBA) [1], dominant pruning (DP) [6].

3.1 Counter Based

The method is executed locally on every node in the network. It has two parameters: T_{RAD} and C. When new packet is received, time $T = (0..T_{RAD}]$ is drawn. Within T the packet counter c is incremented when duplicates of the packet are received. Then, if $c < C$, the packet is retransmitted.

As the method works locally it has very low overhead on additional communication (depends on parameters) and can cope with dynamic changes in the topology (e.g. mobile nodes). The drawback is that the method doesn't guarantee the full network coverage and may select forwarders in such a way, that part of the network will not receive traffic. The C and T_{RAD} parameters can by adjusted. The bigger C leads to better network cover, but also to more forwarders and more messages duplicates. If $C = \infty$ (practically "large enough") the algorithm works as flooding. Bigger T_{RAD} also leads to better coverage but also increase the time of message delivery. The method doesn't assume to create the logical topology, because the decision on packet retransmission can be taken after receiving each packet, but it leads to decreasing the transmission delays. In the evaluation we used the counter based methods to select the forwarders in the initialization phase only. The first choose of each node to retransmit the packet results in selecting that node as one of forwarders (Fig. 2).

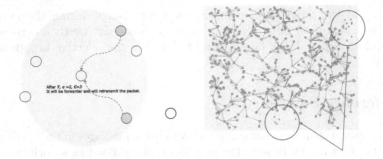

Fig. 2. (a) The counter based method used to packet retransmission, **(b)** The example of logical topology created using CB method, with some unconnected nodes

3.2   Scalable Broadcast Algorithm (SBA)

The algorithm works locally and assumes that every node knows its direct (1-hop) neighbour list. It uses one parameter T_{RAD}. When new broadcast packet is received, a time $T = (0..T_{RAD}]$ is drawn. Every packet header contains sender's neighbours list. Receiver analyses packets, incoming within time T. After T, if there are still nodes in the range that not received packets, the node forward a packet. 100 % cover is guaranteed and the algorithm exhibit good scalability properties as the network size increases. Similarly as CB in the evaluations we assumed the initial phase, in which the first decision of forwarders selection is saved and used for next transmissions. The characteristic of the SBA method is the necessity to transmit the list of neighbours, thus the overhead increase in compare to the CB.

3.3   Dominant Pruning Method

The method utilizes 2-hop neighbourhood information to reduce redundant transmissions. A forwarder, knowing the full 2-hop topology, selects the set of next forwarders among its 1 hop neighbours, to achieve the full cover of all nodes within 2-hop range. Then all designated forwarders repeat that step. This method is called DP local. The forwarders selection is solved as a minimal covering set problem. The optimal solution is a NP-complete problem (N! combinations to check), but the amount of nodes to analyse is usually small. 100 % network cover is guaranteed. The disadvantage is that in relatively large number of forwarders. The overhead on communication is relatively big (necessity to send the list of 2hop neighbours) (Fig. 3).

The method can be also considered as local, but the synchronization is needed. It can be implemented using a token to assure that only one forwarder is able to perform the selection operation The DP method can be implemented using recursive selection of forwarders (DP deep). The forwarder is selected, which has the largest coverage. It sets its best forwarder, and so on (deep selection). When full cover is achieved the decision goes back to the forwarder on

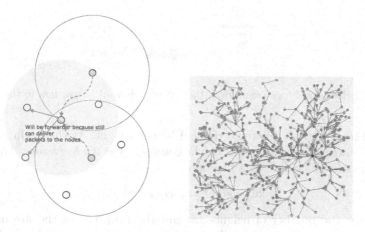

Fig. 3. (a) The Scalable Broadcast Algorithm used to select a forwarder node, (b) The example of logical topology created using SBA method

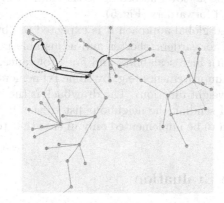

Fig. 4. Deep selection of forwarders in DP method

higher level. As the result the less number of forwarders is achieved but the patches from first node to subsequent nodes (first forwarder) are longer (Fig. 4).

4 The Global Queue Pruning Method

As the stable physical topology in the long term was assumed and the known, designated control point is selected, we decide to propose the new, global approach. It was expected to have significantly "better" topology at the expense of the communication cost in the initialization phase. We propose a novel method, called Global queue pruning (GQP). It is based on dominant pruning, but the designation of forwarders is global (done e.g. by a server or central node) and is based on a queue of potential forwarders. In the initial phase every node sends to the known, control node the list of its 1hop neighbours. The global queue of

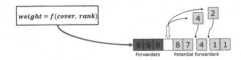

Fig. 5. Selection of forwarders from the list of potential forwarders in the queue

potential forwarders is created, arranged by the weight. At the beginning every node is a potential forwarder as it can be considered as the forwarder. The weight in the queue is calculated as a function:

$$[\text{ht}] \; weight = f(cover, rank)$$

Cover is the number of neighbours and the rank means the distance from the central node. The node with greatest weight value is designated as forwarder. Selection of a forwarder influences on the nodes in the queue (queue is rearranged) by reducing its cover according to the number of neighbours covered by the already selected forwarders (Fig. 5).

Using the presented global approach it is expected to obtain 100 % coverage with lower number of forwarders, shorter and adjustable paths (by influencing on the weight function), high scalability and fault tolerance. The algorithm has also the potential for improvements (e.g. by some refinement phases, and developing more complex weight function). The drawback is the high communication overhead (necessity of sending the neighbour list to the designated node), thus the algorithm is worth to be implemented only in case that topology is relatively stable.

5   Performance Evaluation

The evaluation aim is to compare the reference algorithms (CB, SBA, DP) to the newly developed GQP and to compare the strategy of local and global designation of forwarders (efficiency, fault tolerance, scalability and cost). We used the topology generator described in [7]. The generator includes also DES simulator, statistics, logs and the support for the automatization of evaluations. The methodology is as follow:

1. The generator generates physical topology (random distribution of nodes, but subsequent nodes were located randomly, but within the range of existing nodes, what theoretically guarantee the connectivity between nodes)
2. Based on the physical topology an algorithm was run to designate forwarders. Thus the logical topology was created (in a form of logical tree)
3. The broadcast communication (from central, designated node to all nodes) was simulated to obtain a result for a single broadcast communication. The simulation phase was necessary because the communication during broadcast is possible along the paths different than according to the logical path in the tree. Nodes can receive duplicates e.g. in case if there are in the range of two or more forwarders.

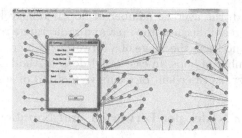

Fig. 6. The example of topology. The purple lines indicates the logical topology, the pink lines indicates the physical connections (Color figure online)

The area of N×N m was analysed. The parameters were: N, number of nodes K, minimal distance between nodes Dmin, maximum distance Dmax, radio range R. It was also possible to adjust the average number of neighbours Navg. In such case the Dmax parameter was calculated automatically. The assumed parameters were: $N = 1000\,m$, $K = 100..500$, D_{min} : 5 m, Node Range: 200 m.

All described above algorithms were evaluated (CB, SBA, DPlocal, DPdeep, GQP). For each of them 200 simulations were carried out (for different physical topologies). Results present the averages (Fig. 6).

5.1   The Average Number of Hops

The number of hops is an important parameters that influences of the delays in communications, and especially in ad-hoc or grid network on the energy consumption (the longer the path are, the more resources are used by the intermediate node to deliver message. The results are presented on Fig. 7.

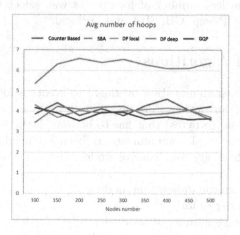

Fig. 7. Average number of hops as a function of total number of nodes

As it is presented, the average number of hop is relatively stable while the number of nodes increase, because the area and radio range remains unchanged. Only the number of nodes in the range of a forwarder is increasing, what doesn't influence on the number of hops. In case of DPdeep method the number of hops is significantly higher, as the result of recursive method of selecting forwarders.

5.2   The Number Nodes per Forwarder and Number of Forwarders

Generally the lower the number of forwarders is, the more optimal logical topology is created. Less forwarders generate smaller communication overhead, less number of duplicates etc. The figures below present two results: the number of forwarders and number of nodes within the range of a forwarder (Fig. 8).

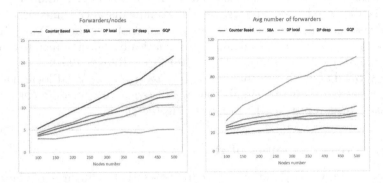

Fig. 8. The average number of nodes within the range of a forwarder and the number of forwarders in a function of number of nodes

As presented, the less number of forwarders was selected in case of GQP method, than SBA, DPdeep, CB, SBA and DPlocal.

6   The Cost of Algorithms

The cost reflects the communication overhead to create a logical topology and designate the set of forwarders. The calculated value is proportional to the amount of information (in bytes) that has to be sent in the initialization phase. The cost includes the local communication (between neighbours) and global communication with designated control node. The calculations includes the parameters:

a	information about one node
f	number of forwarders
$neigh$	average number of forwarders
ntf	number of nodes per forwarder
n	total number of nodes

In case of analysed algorithms the cost can be expressed as follows:

Counter based: $cost = a(f + n)$
SBA, DPlocal, DPdeep: $cost = a(f * neigh + n)$
GQP: $cost = a(n * neigh + f)$

The Fig. 9 presents the comparison of costs:

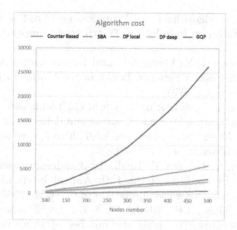

Fig. 9. The comparison of algorithms costs

As it is presented the cost of GQP algorithms is significantly greater than in all remaining methods and it grows geometrically with the number of nodes.

7   Conclusions

The proposed Global Queue Pruning GQP algorithm creates the logical topology that consists of considerably lower number of forwarding nodes in comparison to the three other commonly used methods, evaluated in the paper: Counter Based, Scalable Broadcast and Dominant Pruning. The paths generated by the GQP are relatively short and guarantee the delivery to all the nodes in the network. The important drawback is the communication cost to create the topology in the initialisation phase. In case of stable physical topology and the communication based on one designated control node the GQP algorithm is worth to be considered.

References

1. Boukerche, A. (ed.): Algorithms and Protocols for Wireless and Mobile Ad Hoc Networks. Wiley Series on Parallel and Distributed Computing. Wiley, New York (2009)
2. Garey, M.R., Johnson, D.S.: Computers and Intractability: A Guide to the Theory of NP-Completeness. A Series of Books in the Mathematical Sciences. W. H. Freeman, San Francisco (1979)
3. Izumi, S., Matsuda, T., Kawaguchi, H., Ohta, C., Yoshimoto, M.: Improvement of counter-based broadcasting by random assessment delay extension for wireless sensor networks, pp. 76–81. IEEE, October 2007. http://ieeexplore.ieee.org/lpdocs/epic03/wrapper.htm?arnumber=4394901
4. Keshavarz-Haddad, A., Ribeiro, V., Riedi, R.: Broadcast capacity in multihop wireless networks. In: Proceedings of the 12th Annual International Conference on Mobile Computing and Networking, pp. 239–250. ACM (2006)
5. Khabbazian, M., Blake, I.F., Bhargava, V.K.: Local broadcast algorithms in wireless ad hoc networks: reducing the number of transmissions. IEEE Trans. Mobile Comput. **11**(3), 402–413 (2012). http://ieeexplore.ieee.org/lpdocs/epic03/wrapper.htm?arnumber=5740910
6. Lim, H., Kim, C.: Flooding in wireless ad hoc networks. Comput. Commun. **24**(3–4), 353–363 (2001). http://linkinghub.elsevier.com/retrieve/pii/S0140366400002334
7. Nowak, S., Nowak, M., Grochla, K.: MAGANET – on the need of realistic topologies for AMI network simulations. In: Kwiecień, A., Gaj, P., Stera, P. (eds.) CN 2014. CCIS, vol. 431, pp. 79–88. Springer, Heidelberg (2014)
8. Nowak, S., Nowak, M., Grochla, K.: Properties of advanced metering infrastructure networks' topologies. In: 2014 IEEE Network Operations and Management Symposium (NOMS), pp. 1–6. IEEE (2014)
9. Tanenbaum, A.S., Wetherall, D.: Computer Networks, 5th edn. Pearson Prentice Hall, Boston (2011)
10. Yi, J., Clausen, T., Igarashi, Y.: Evaluation of routing protocol for low power and Lossy Networks: LOADng and RPL, pp. 19–24. IEEE, December 2013. http://ieeexplore.ieee.org/lpdocs/epic03/wrapper.htm?arnumber=6728773

Network Layer Benchmarking: Investigation of AODV Dependability

Maroua Belkneni[1]([⊠]), M. Taha Bennani[1,2], Samir Ben Ahmed[1,2],
and Ali Kalakech[3]

[1] University of Tunis El Manar, Tunis, Tunisia
belknenimaroua@gmail.com
[2] University of Carthage, Carthage, Tunisia
taha.bennani@enit.rnu.tn, samir.benahmed@fst.rnu.tn
[3] Lebanese University, Beirut, Lebanon
akalakech@ul.edu.lb

Abstract. In wireless sensor networks (WSN), the sensor nodes have a
limited transmission range and storage capabilities as well as their energy
resources are also limited. Routing protocols for WSN are responsible
for maintaining the routes in the network and have to ensure reliable
multi-hop communication under these conditions. This paper defines the
essential components of the network layer benchmark, which are: the
target, the measures and the execution profile. This work investigates the
behavior of the Ad Hoc On-Demand Distance Vector (AODV) routing
protocol in situations of link failure. The test bed implementation and
the dependability measures are carried out through the NS-3 simulator.

1 Introduction

Wireless Sensor Networks (WSNs) represent a concrete solution for building
next-generation critical monitoring systems with reduced development, deployment, and maintenance costs [3]. WSNs applications are used to perform many
critical tasks. Properties that such applications must have include availability,
reliability, security and etc. The notion of dependability captures these concerns
within a single conceptual framework, making it possible to approach the different requirements of a critical system in a unified way. The unique characteristics
of WSNs applications make dependability satisfaction in these applications more
and more significant [8].

The structure of the paper is as follows. In Sect. 2, we show the related
work. In Sect. 3, we describe the benchmark target. Next, in Sect. 4, is held the
execution profile. Section 5 defines the faultload specification. Section 6 describes
measurements and simulation results. Finally, Sect. 7 concludes the paper.

2 Related Work

Various routing protocols have been compared, in the literature, using different
aspects, namely the evaluation of performance or dependability. In the first case,

© The Author(s) 2016
T. Czachórski et al. (Eds.): ISCIS 2016, CCIS 659, pp. 225–232, 2016.
DOI: 10.1007/978-3-319-47217-1_24

a set of measures is usually used to compare different solutions. Authors in [7] describe a number of quantitative parameters that can be used to evaluate the performance of Mobile Ad hoc Networking (i.e. MANET) routing protocols. In contrast the dependability measures define many properties like: time-to-failure and time-to-recovery [4]. Other measures may define the network and the sensing reliability. To perform such analysis we can use approaches like: simulation, emulation and real-world experiments [9]. We aim to define a fault injection based evaluator that handle errors and analyze the sensor networks reliability [1].

3   Benchmark Target

The network layer provides various types of communications. Which are not only messages delivering and the network layers yielded notification, but, also the paths discovery and its maintenance. Therefore, these two services are mandatory to build the workload that assesses the network layer dependability. We have used AODV [5] as the reference protocol to simulate these two services using NS3 [6].

Route Calculation: AODV broadcasts a Route Request (RREQ) to all its neighbors. Then it propagates the RREQ through the network, unless, it reaches either the destination or the node holding the newest route to the destination. The destination node sends back a RREP response to the source to prove the validity of the route [2]. Route Reply (RREP) message is unicast back and it contains hop_count, dest_ip address, dest_seqno, src_ip address and lifetime as shown in Fig. 1.

Hop count	Dest Address	DSN	Source Address	Lifetime

Fig. 1. RREP packet format

Route Maintenance: AODV sends these broadcasted "hello" messages (a special RREP) which are simple protocols used by the neighbors to refresh their valid routes set. If one node no longer receives the hello messages from a particular node, it deletes all the routes that use the unreachable link, and that form the set of the valid routes. It also notifies the affected set of nodes by sending to them a link failure notification (a special RREP see Fig. 2).

DestCount	Unreachable Dest Address	Unreachable DSN

Fig. 2. RERR packet format

The forwardup() operation of processes, a protocol data unit (PDU) messages and delivers it to the upper layers, whereas the Receive() operation provides the requests response. These two activities define services offered by the LLC Layer.

4   Execution Profile

The execution profile activates the target system with either a realistic or a synthetic workload. Unlike performance benchmarking, which includes only the workload, the dependability assessment also needs the definition of the faultload. In this section, we describe the structure and the behavior of the workload.

4.1   Workload Structure

To apply our approach to a real structure, we chose to monitor the stability of a bridge. Figure 3 introduces the topology of the nodes which is a 3D one. In our experiments, we vary the number of nodes within the range of 10 to 50 (see Table 1). The more we define nodes, the more is dependable the structure. With ten nodes, the structure has one redundant path between the source node and the sink. Then, even though one node had failed, the emitter node would have transmitted a packet to the sink. When the structure has more nodes, it will tolerate more than one node failure.

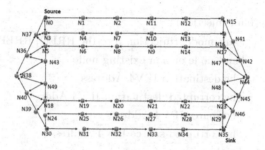

Fig. 3. Scheme of the considered bridge and resulting topology

Table 1. Simulation parameters

Network Simulator	NS3
Channel type	Channel/Wireless channel
MAC type	Mac/802.11
Routing Protocol	AODV
Simulation Time	100 s
Number of Nodes	10, 20, 30, 40, 50
Data payload	512 bytes
Initial energy	10 J

4.2   Workload Behavior

As the assessed services is the route establishment and its maintenance by the network protocol, our workload consists on the sending of a packet from a source to the sink node. The Table 1 below summarises the simulations' parameters.

5   Faultload Specification

It would be awkward to identify the origin of the failure using multiple modifications, therefore, to avoid the correlation drawback, our benchmark assesses the WSN behavior using a single fault injection. As the source node triggers the communication, the route construction and its maintenance, we will inject faults within the packets received by this node and therefore the change in field of its routing table. Since the source node receives the RREP packets in the route identification phase and RERR in the maintenance one, we will inject into its different fields, described in the Table 2 below.

Table 2. The variable declaration

Fixed variable (fault injection)	
F_model	Fault model (injection into the RREQ, RREP or RRER)
F_type:	Fault node or non existing node
Dest:	The destination IPV4 Address
Cptd_Dest:	The corrupted destination IPV4 Address
SRC:	The source IPV4 Address
Cptd_SRC:	The corrupted source IPV4 Address
HC:	The hop count
Cptd_HC:	The corrupted hop count
LF:	The life time
Cptd_LF:	The life time
DSN:	The destination sequence number
Cptd_DSN:	The corrupted destination sequence number
UNDest:	Unreachable Dest Address
UNDSN:	Unreachable DSN
Control function	
SetDst():	Set destination address
SetDstSeqno():	Set destination sequence number
SetHopCount():	Set hop count
SetOrigin():	Set source address

The table above introduces two set of elements: Fixed variables and control functions which are mandatory to specify the faultload. Fixed variables are the elementary parameters of the fault, they identify the packet's fields and their relative corrupted values. Also, the fault model specifies the faulty packet which could be the RREP or RERR packet and the fault type initializes the node's address using a random value belonging to the network or an imaginary one. All these values have to stay constant during one the simulation. The functions, belonging to the "Control functions", change the fields of control packets.

The CTL (Computation Tree Logic) formulae written below specify the fault-load used to assess the dependability of the routing layer. The expression (1) and (5) specifies respectively, a fault injection within the RREP and RERR packet. The fault type can take a false value of an another node within our architecture or a value of a non existing one. When we inject in the RREP packet, the fault may cover four fields: HC(3), DST(3), SRC(4) or DSN(4). In the RERR injection, the fault may alter these following fields: UNDST, UNDSN(7). In this section, we present the fault injection specification in the AODV protocol. The fault injection will be modeled in the primitive Forwardup () at the entrance of the network layer.

RREP Injection:

$$Fault\_model = RREP \land \tag{1}$$
$$(Fault\_type = fault \lor non\_existing) \land \tag{2}$$
$$(DST = Cptd\_DST \lor HC = Cptd\_HC \lor \tag{3}$$
$$SRC = Cptd\_SRC \lor DSN = Cptd\_DSN \lor LF = Cptd\_LF) \tag{4}$$

RERR Injection:

$$(Fault\_model = RERR \land \tag{5}$$
$$(Fault\_type = fault \lor non\_existing) \land \tag{6}$$
$$(UNDST = Cptd\_DST \lor UNDSN = Cptd\_DSN)) \tag{7}$$

6   Measurements and Simulation Results

We need measurements to determine the dependability of the WSN:

- Remaining energy: Is the average of remaining energy of all nodes.
- Time of route recovery: It is the time taken by a protocol to find another path to the destination.
- Time of route identification: It is the time taken by a protocol to find a route to the destination.

6.1   Route Calculation

In the following sections, we will present the results and analyze them. The after simulation results are viewed in the form of line graphs. The study of AODV is

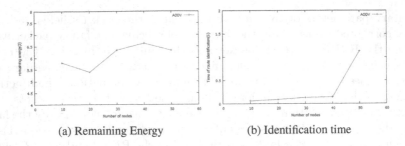

(a) Remaining Energy   (b) Identification time

Fig. 4. Fault free simulation

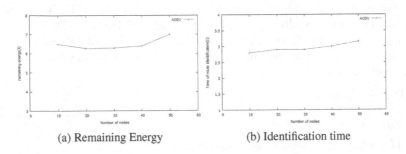

(a) Remaining Energy   (b) Identification time

Fig. 5. Fault injection simulation of AODV

based on the varying of the workload and the faultload. This study is done on parameters remaining energy and time of route identification. The Fig. 4a shows the AODV power consumption compared to the number of nodes. In the Fig. 4b, we note that AODV is very fast to find the route especially when the number of nodes decreases.

The AODV protocol is robust to the hopcount and the lifetime fields injection. It find the route and keep the same performances as if we did not interfere.

AODV is not robust to the source address fields injection. When we inject in a node that belongs to the route and despite that there is an another one, the protocol don't find the path. With the Dest and the DSN fields injection, the protocol sends another RREQ which increases the route identification time and the remaining energy as shown in Fig. 5.

6.2   Route Maintenance

To evaluate the route maintenance we produce the failure of an intermediate node. Figure 6 shows the remaining energy and the recovery time without fault injection. To study the behavior of the AODV protocol during the route mainte-nance, we injected the fault after provoking the failure of the intermediate node. The fault model and the injection model used are defined in the section four. AODV protocol is robust with respect to the both filds to the Unreachable Dest Address and Unreachable DSN. Nevertheless the RERR packet rate increases which saves energy during the simulation.

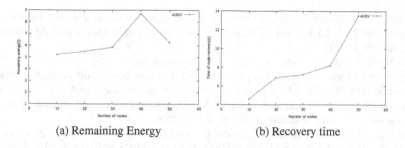

(a) Remaining Energy   (b) Recovery time

Fig. 6. Fault free simulation

7 Conclusion

We studied the AODV dependability, considering the remaining energy, the time of route recovery and the time of route identification. After the benchmarking campaigns, we noticed that the AODV protocol is robust with respect to eight filds introduced in the section three except the source address in the packet RREP.

References

1. Sailhan, F., Delot, T., Pathak, A., Puech, A., Roy, M.: Dependable Sensor Networks, Atelier sur la GEstion des Donnes dans les Systmes d'Information Pervasifs (GEDSIP) au sein de la confrence INFormatique des ORganisations et Systmes d'Information et de Dcision (INFORSID), pp. 1–15, May 2010
2. Kumari, S., Maakar, S., Kumar, S., Rathy, R.K.: Traffic pattern based performance comparison of AODV, DSDV and OLSR MANET routing protocols using freeway mobility model. Int. J. Comput. Sci. Inf. Technol. **2**, 1606–1611 (2011)
3. Akyildiz, I.F., Su, W., Sankarasubramaniam, Y., Cayirci, E.: Wireless sensor networks: a survey. Comput. Netw. **38**(4), 393–422 (2002)
4. Chipara, O., Lu, C., Bailey, T.C., Roman, G.-C., Networks, reliable clinical monitoring using wireless sensor: experiences in a step-down Hospital unit. In: Proceedings of the 8th ACM Conference on Embedded Networked Sensor Systems, vol. 14, pp. 155–168 (2010)

5. Perkins, C.E., Royer, E.M.: Ad-hoc on demand distance vector routing. In: Proceedings of the 2nd IEEE Workshop on Mobile Computing Systems and Applications, pp. 90–100 (1999)
6. The NS-3 Network Simulator. http://www.nsnam.org
7. Corson, S., Macker, J.: Routing protocol performance issues and evaluation considerations. RFC2501, IETF Network Working Group, January 1999
8. Taherkordi, A., Taleghan, M.A., Sharifi, M.: Dependability considerations in wireless sensor networks applications. J. Netw. 1(6) (2006)
9. Kulla, E., Ikeda, M., Barolli, L., Xhafa, F., Younas, M., Takizawa, M.: Investigation of AODV throughput considering RREQ, RREP and RERR packets. In: Advanced Information Networking and Applications (AINA), pp. 169–174 (2013)

Occupancy Detection for Building Emergency Management Using BLE Beacons

Avgoustinos Filippoupolitis$^{(\boxtimes)}$, William Oliff, and George Loukas

Department of Computing and Information Systems,
University of Greenwich, London, UK
{a.filippoupolitis,w.oliff,g.loukas}@gre.ac.uk

Abstract. Being able to reliable estimate the occupancy of areas inside a building can prove beneficial for managing an emergency situation, as it allows for more efficient allocation of resources such as emergency personnel. In indoor environments, however, occupancy detection can be a very challenging task. A solution to this can be provided by the use of Bluetooth Low Energy (BLE) beacons installed in the building. In this work we evaluate the performance of a BLE based occupancy detection system geared towards emergency situations that take place inside buildings. The system is composed of BLE beacons installed inside the building, a mobile application installed on occupants' mobile phones and a remote control server. Our approach does not require any processing to take place on the occupants' mobile phones, since the occupancy detection is based on a classifier installed on the remote server. Our real-world experiments indicated that the system can provide high classification accuracy for different numbers of installed beacons and occupant movement patterns.

1 Introduction

Thanks to its exceptionally low power requirements, low cost and compatibility with most mobile devices and computers, Bluetooth low energy (BLE) is rapidly proving to be a very practical technology in e-health, sports, fitness, marketing in malls and other applications. We argue that its ability to provide proximity information with sufficient accuracy can extend its use in emergency management too, especially in buildings and other confined spaces, where traditional localisation technologies often fail. For example, having a mechanism to estimate the occupancy of different areas within a building can help emergency personnel produce a more optimal plan of action. In the literature on emergency management supporting technologies, it is often assumed that the emergency personnel or unmanned technical systems involved are aware of the locations where there are individuals requiring assistance/rescue [5,6,12], but this assumption can be highly inaccurate in many real-life situations. For example, during the 2015 terrorist attack in a Tunis museum, two tourists spent the night hiding in the museum only to be found the next day. Afraid to attract the attention of the terrorists, they had refrained from using their phones. BLE can

© The Author(s) 2016
T. Czachórski et al. (Eds.): ISCIS 2016, CCIS 659, pp. 233–240, 2016.
DOI: 10.1007/978-3-319-47217-1_25

help both occupancy detection and indoor localisation, as has been acknowledged in a US Federal Communications Commission roadmap for BLE use in conjunction with WiFi to help locate 911 callers inside buildings.

There is a wide range of BLE based applications targeted to building occupants, including indoor navigation [7], activity recognition [1] and remote healthcare monitoring [11]. With respect to indoor occupancy estimation and localisation, we can find various approaches targeting different area types. The authors in [4] discuss the use of Apple's iBeacon protocol for building occupancy detection. They evaluated their approach using a single room and predicted whether the occupant was inside or outside. A system that detects the locations of occupants inside an office is presented in [2]. This is used to control a building management system that the authors evaluate inside an office area. The estimation of a building's occupancy using Arduino based beacons is described in [3]. The authors evaluate the system by estimating an occupant's presence inside or outside a single room. The authors in [9] employ iBeacons inside the floor of a building in order to evaluate the performance of an occupancy estimation system for hospitals. Their system has a high overall accuracy but there are no accuracy results for individual areas. In [10] the authors propose an indoor localisation system that uses BLE beacons inside an office building. Their approach achieves a high localisation accuracy (for 75 % of the time the localisation error is lower than 1.8 m) however they have not evaluated the effect of walking speed or beacon locations. Finally, the authors in [8] propose an indoor localisation system based on BLE beacons. The system is evaluated inside a single room and although they claim a high accuracy rate, their results are limited.

Our approach is targeted towards emergency situations and aims to provide an estimate of the number of occupants inside areas such as offices, laboratories and conference rooms. Even if our proposed system stops functioning (e.g., due to a natural or man-made disaster), it is still able to provide very useful information related to the spatial distribution of the occupants at the time before the incident took place.

2   Description of the System

Our approach is based on the use of BLE beacons located inside the building that communicate with a mobile application installed on the occupant's phone. The beacons use a non-connectible mode, the BLE advertising mode, to periodically broadcast advertisement packets that include information such as the beacon's unique ID. A mobile phone located in the vicinity of a beacon receives the packets and processes them using a mobile application. In a commercial setting the main assumption is that the mobile application has knowledge of the beacons' location inside the building and of the mapping between beacons and rooms or areas. This information is then used by the mobile application, in conjunction with the received BLE packets, in order to calculate the user's location inside the building. Finally, the mobile application sends its location to a remote server which then replies with contextual information (such as a targeted micro-location based advertisement).

Fig. 1. System architecture

Figure 1 illustrates our system's operation in an emergency situation that takes place inside a building. The mobile application installed on the occupants' phones receives BLE messages from multiple beacons. It then sends their RSSI values and respective beacon IDs to the remote control server. Finally the server, upon reception of this information from a mobile device, uses a trained classifier to update the building occupancy estimation. Our approach has numerous advantages. Firstly, the mobile phone does not need to know the mapping between beacon ID and location of beacons inside the building. Also, the mobile phone does not process the received beacon packets to calculate its location and the remote control server does not send information back to the mobile phone. Since our system does not involve localisation related processing by the mobile application, we can use mobile devices that have low computational power and memory capacity. The remote control server is responsible for processing the data that the mobile application sends and for calculating the building occupancy. To achieve this, we conduct a single data gathering phase during which the data gathered are used to train a classifier. Section 3 provides further details on this process. After the data gathering phase has been completed, the system is able to operate in normal mode as shown in Fig. 1.

For our BLE beacons we used a Raspberry Pi 2 with a Bluetooth 4 BLE USB module. We implemented the iBeacon protocol, which is the BLE beacon implementation proposed by Apple. By using an open platform such as the Raspberry Pi, we avoided the limitation of being tied to a specific beacon manufacturer. To identify the iBeacons, we used a Universally Unique Identifier (UUID), a major number and a minor number for each of them. The UUID is used to separate the iBeacons being used in our experiments from other unassociated Bluetooth devices. The major number is used to define local groups of iBeacons (e.g. belonging to a certain building or floor) and the minor number is used to define each individual iBeacon within a local group. We can use our Android mobile application for the data gathering phase as well as for the normal operation of the system. When the mobile application receives a BLE advertising data packet from an iBeacon during the data gathering phase, it extracts and logs the UUID, major number, minor number and transmission (Tx) power of the beacon from the packet's payload. The application also logs the received signal strength indicator (RSSI) of each received BLE packet. Finally, an area label is manually assigned to each packet by the user based on his actual location inside the building. Under normal operation mode, the mobile application simply receives BLE packets from beacons and sends their RSSI values and respective beacon IDs to the server. The remote control server processes the data sent by the mobile application in order to calculate the occupancy of the building. In normal operation mode, the server receives information from the mobile application running on an occupant's mobile phone and uses a trained classifier to update the building occupancy estimation. The training of this classifier is performed during the initial data gathering phase. We must note, however, that it is not necessary for the training to take place in the server. The only requirement is that the trained classifier model is stored on the server so that it can be used during normal operation.

3   Experimental Evaluation

We evaluated the performance of our system in the computer laboratory of the University of Greenwich. This is essentially an office space that includes objects such as desks, benches, computers, panels and chairs. We have identified five areas inside the laboratory (A1-A5), as illustrated in Fig. 2. An orthogonal grid was used to map the experimental area, with each grid square equal to an area of $1\,m^2$. We investigated two beacon deployment configurations: one involving four beacons and one involving seven beacons, as shown in Figs. 2(a)–(b). For the data gathering phase, we used our mobile application in data gathering mode. The beacons' transmission frequency was set to 8 Hz and their transmission power to 4 dBm. To increase the level of realism, instead of standing inside each area we moved according to a "Walk and Stop" pattern that involved spending 10 s on each grid point before moving to the next one. For each BLE packet received the mobile application logged the UUID, major number, minor number and RSSI

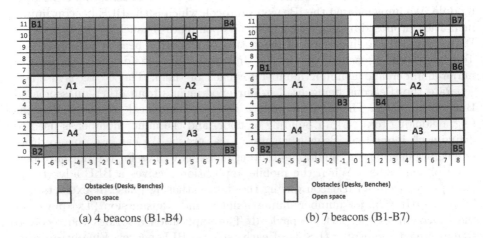

(a) 4 beacons (B1-B4)   (b) 7 beacons (B1-B7)

Fig. 2. Experimental area and beacon positions for the two different configurations

and we assigned an area label (A1 to A5) based on our actual location. For each of the two beacon setups, we conducted two runs of this data gathering phase. This resulted in a dataset size of over 44,000 packets for the 4 beacon setup and of over 78,000 packets for the 7 beacon setup.

We modelled our problem as a multi-class classification problem, with the number of classes equal to the number of areas in our environment (i.e. five classes). Our raw dataset contained individual packets coming from specific beacon IDs, with a respective RSSI value and an area label. To transform this to a dataset that can be used to train a classifier, we used a data segmentation approach involving a non-overlapping sliding window. For each beacon inside a specific area, we calculated the average and the standard deviation of its RSSI over the window samples and used these as the features of our classification problem. For the four beacon setup, this resulted in eight features while for the seven beacon setup we had fourteen features. For our classifier we have chosen a support vector machine with radial basis function kernel (SVM). The reason behind this choice is that SVMs can successfully deal with non-linearly separable data. We partitioned the dataset into 80 % training set and 20 % test set and used 10-fold cross validation for hyper-parameter tuning. We used a confusion matrix for presenting our classification results, where its rows represent the instances in an actual class and its columns the instances in a predicted class. The values of the matrices are normalised by the number of elements in each class, to better illustrate the classification accuracy for each class.

3.1   Results for "Walk and Stop" Scenario

Figure 3 illustrates our classification results for the "Walk and Stop" scenario and the four beacons setup. In the case of a 0.5 s window, we can observe that the classification accuracy ranges from 64 % to 89 %. Increasing the window size to 1 s, as depicted in Fig. 3(b), improves the classification performance especially for Area 2 where its classification accuracy has now increased from 64 % to 81 %. Further increasing the window size to 2 s, as shown in Fig. 3(c), does not provide a clear improvement of the classification accuracy. For example, although Area 1 is now classified with 100 % accuracy, the performance of the classifier for Area 2 has dropped to 68 %. By inspecting Figs. 3(a)–(c) we can observe a consistently low performance of our classifier with respect to Area 2. This can be explained if we look at the spatial distribution of beacons with respect to areas, as depicted in Fig. 2(a). We can observe that the number of beacons is less than the number of areas (four versus five respectively). Moreover, each Area can be associated with one specific beacon which is closest to it: Area 1 with Beacon 1, Area 4 with Beacon 2, Area 5 with Beacon 4 and Area 3 with Beacon 3. However, there is no one Beacon that can be associated with Area 2. The two closest beacons to Area 2 are Beacon 4 and Beacon 3. This sparse beacon deployment explains the low classification performance for Area 2. We can also verify from Figs. 3(a)–(c) that Area 2 is consistently misclassified as Area 3 or Area 5, which are the two areas closest to Beacon 3 and Beacon 4.

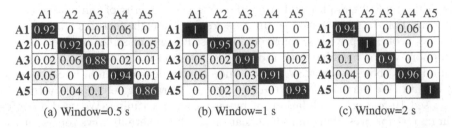

	A1	A2	A3	A4	A5
A1	0.82	0	0.04	0.13	0.01
A2	0	0.64	0.17	0.01	0.17
A3	0.03	0.11	0.8	0.05	0
A4	0.16	0.01	0.01	0.81	0
A5	0.04	0.07	0	0	0.89

(a) Window=0.5 s

	A1	A2	A3	A4	A5
A1	0.86	0.02	0	0.04	0.09
A2	0	0.81	0.12	0	0.06
A3	0.02	0.22	0.76	0	0
A4	0.12	0.03	0	0.82	0.03
A5	0.03	0.11	0	0	0.86

(b) Window=1 s

	A1	A2	A3	A4	A5
A1	1	0	0	0	0
A2	0	0.68	0.16	0	0.16
A3	0.04	0.24	0.72	0	0
A4	0.09	0	0	0.91	0
A5	0.05	0.05	0	0	0.9

(c) Window=2 s

Fig. 3. Confusion matrices for SVM, using 4 beacons and different window sizes ("Walk and Stop" Scenario)

	A1	A2	A3	A4	A5
A1	0.92	0	0.01	0.06	0
A2	0.01	0.92	0.01	0	0.05
A3	0.02	0.06	0.88	0.02	0.01
A4	0.05	0	0	0.94	0.01
A5	0	0.04	0.1	0	0.86

(a) Window=0.5 s

	A1	A2	A3	A4	A5
A1	1	0	0	0	0
A2	0	0.95	0.05	0	0
A3	0.05	0.02	0.91	0	0.02
A4	0.06	0	0.03	0.91	0
A5	0	0.02	0.05	0	0.93

(b) Window=1 s

	A1	A2	A3	A4	A5
A1	0.94	0	0	0.06	0
A2	0	1	0	0	0
A3	0.1	0	0.9	0	0
A4	0.04	0	0	0.96	0
A5	0	0	0	0	1

(c) Window=2 s

Fig. 4. Confusion matrices for SVM, using 7 beacons and different window sizes ("Walk and Stop" Scenario)

By increasing the number of beacons to seven, we observed a significant improvement in the classification accuracy for all window sizes, as depicted in Fig. 4. For a window size of 0.5 s the classification accuracy ranges from 86 % to 94 %, as shown in Fig. 4(a). Figure 4(b) illustrates the results for a window size equal to 1 s. We can verify that increasing the window size improves the classification accuracy, which now ranges from 91 % to 100 %. Finally, further increasing the window size to 2 s does not yield a significant improvement in accuracy, as Fig. 4(c) shows. We should also note that in the seven beacon configuration we do not observe the consistent misclassification of Area 2, as was the case in the four beacon configuration.

3.2 Results for "Random Walk" Scenario

To investigate the effect of the movement pattern on the classification accuracy, we have conducted an additional experiment with the seven beacon configuration. This time, we moved inside each area without stopping on grid points. The movement involved randomly choosing a destination grid square point within each area, walking towards it, then choosing another one and repeating the same procedure for each area. The total duration of this "Random Walk" scenario was equal to that of the "Stop and Walk" scenario for the seven beacon configuration, in order to achieve the same dataset size.

As we can observe from Fig. 5, the classification accuracy is lower compared to the one shown in Fig. 4. For a window size of 0.5 s, the accuracy ranges

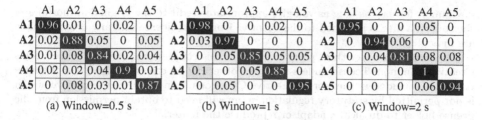

Fig. 5. Confusion matrices for SVM, using 7 beacons and different window sizes ("Random Walk" Scenario)

between 84 % and 96 %. Increasing the window size from 0.5 s to 1 s results in an improvement in accuracy which ranges between 85 % and 97 %. A window size of 2 s improves the classification accuracy further, especially for Area 4 which increases to 100 % from the 85 % of the 1 s window case.

This was expected, as the constant movement of the occupant in the "Random Walk" makes training the system more challenging, resulting in reduced accuracy compared to the more static "Walk and Stop" case. At the same time, increasing the size of the window results in averaging RSSI values over a longer time interval for each data point. This compensates for the constant movement of the occupant but reduces the responsiveness of the system, because under normal system operation the server would have to wait for 2 s before receiving RSSI data from the mobile application.

4 Conclusions and Future Work

In this work, we have evaluated the performance of a BLE based occupancy detection system geared towards emergency situations that take place inside buildings. The system is composed of BLE beacons installed inside the building, a mobile application installed on occupants' mobile phones and a remote control server located outside the building. We do not require any localisation calculations to take place on the mobile phone, since the occupancy detection is based on a classifier installed on the remote server. Our real-world experiments indicated that the system can provide a high classification accuracy for different beacon deployment configurations and movement patterns of the building occupants. In future work, we will investigate a greater range of occupant walking speeds and beacon deployment configurations. We also plan to study how our system's performance is affected by different beacon transmission frequencies. Finally, we believe it is worth investigating the use of machine learning algorithms based on neural networks and deep learning to evaluate whether they can further improve the classification accuracy of our system.

medium or format, as long as you give appropriate credit to the original author(s) and the source, a link is provided to the Creative Commons license and any changes made are indicated.

The images or other third party material in this chapter are included in the work's Creative Commons license, unless indicated otherwise in the credit line; if such material is not included in the work's Creative Commons license and the respective action is not permitted by statutory regulation, users will need to obtain permission from the license holder to duplicate, adapt or reproduce the material.

References

1. Alam, M., Pathak, N., Roy, N.: Mobeacon: an iBeacon-assisted smartphone-based real time activity recognition framework. In: Proceedings of the 12th International Conference on Mobile and Ubiquitous Systems: Computing, Networking and Services (2015)
2. Choi, M., Park, W.K., Lee, I.: Smart office energy management system using bluetooth low energy based beacons and a mobile app. In: 2015 IEEE International Conference on Consumer Electronics (ICCE), pp. 501–502. IEEE (2015)
3. Conte, G., De Marchi, M., Nacci, A.A., Rana, V., Sciuto, D.: BlueSentinel: a first approach using iBeacon for an energy efficient occupancy detection system. In: BuildSys@ SenSys, pp. 11–19 (2014)
4. Corna, A., Fontana, L., Nacci, A., Sciuto, D.: Occupancy detection via iBeacon on Android devices for smart building management. In: Proceedings of the 2015 Design, Automation & Test in Europe Conference & Exhibition, pp. 629–632. EDA Consortium (2015)
5. Dimakis, N., Filippoupolitis, A., Gelenbe, E.: Distributed building evacuation simulator for smart emergency management. Comput. J. **53**(9), 1384–1400 (2010)
6. Filippoupolitis, A., Gorbil, G., Gelenbe, E.: Spatial computers for emergency support. Comput. J. **56**(12), 1399–1416 (2012)
7. Fujihara, A., Yanagizawa, T.: Proposing an extended iBeacon system for indoor route guidance. In: 2015 International Conference on Intelligent Networking and Collaborative Systems (INCOS), pp. 31–37. IEEE (2015)
8. Kajioka, S., Mori, T., Uchiya, T., Takumi, I., Matsuo, H.: Experiment of indoor position presumption based on RSSI of Bluetooth LE beacon. In: 2014 IEEE 3rd Global Conference on Consumer Electronics (GCCE), pp. 337–339. IEEE (2014)
9. Lin, X.Y., Ho, T.W., Fang, C.C., Yen, Z.S., Yang, B.J., Lai, F.: A mobile indoor positioning system based on iBeacon technology. In: 2015 37th Annual International Conference of the IEEE Engineering in Medicine and Biology Society (EMBC), pp. 4970–4973. IEEE (2015)
10. Palumbo, F., Barsocchi, P., Chessa, S., Augusto, J.C.: A stigmergic approach to indoor localization using bluetooth low energy beacons. In: 2015 12th IEEE International Conference on Advanced Video and Signal Based Surveillance (AVSS), pp. 1–6. IEEE (2015)
11. Sugino, K., Katayama, S., Niwa, Y., Shiramatsu, S., Ozono, T., Shintani, T.: A bluetooth-based device-free motion detector for a remote elder care support system. In: 2015 IIAI 4th International Congress on Advanced Applied Informatics (IIAI-AAI), pp. 91–96. IEEE (2015)
12. Timotheou, S., Loukas, G.: Autonomous networked robots for the establishment of wireless communication in uncertain emergency response scenarios. In: Proceedings of the 2009 ACM Symposium on Applied Computing, pp. 1171–1175. ACM (2009)

RFID Security: A Method
for Tracking Prevention

Jarosław Bernacki$^{(\boxtimes)}$ and Grzegorz Kołaczek

Department of Computer Science, Wrocław University of Technology,
Wrocław, Poland
{jaroslaw.bernacki,grzegorz.kolaczek}@pwr.edu.pl

Abstract. RFID-tags are very small and low-cost electronic devices
that can store some data. The most popular are passive tags that do not
have own power source, which allows for far-reaching miniaturization.
The primary use of RFID-tags is to replace barcodes. Their industrial
importance is constantly growing because in contrast to barcodes, manual manipulation of the object code is not required. RFID-tags are also
used for detection and identification of objects. This enables tracking of
objects in technological processes. At the moment, the most widespread
use of RFID tags is identification of sold goods. However, the possibility of tracking carries the risk that improper subject can track the tags
and consequently track a person who is in possesion of tagged subject.
Therefore in this paper a method for tracking prevention is considered.

Keywords: Internet of Things · RFID · Privacy protection · Tracking
prevention

1 Introduction

*Internet of Things (IoT) is the convergence of Internet with Radio Frequency
IDentification (RFID), Sensor and smart objects. IoT can be defined as "things
belonging to the Internet" to supply and access all of real-world information* [13].
RFID is said to give rise to the IoT. RFID are systems that consist of three
fundamental elements: *tags, reader* and a database system. Tags (also called
transponders) are "small" electronic devices, highly constrained. They usually
do not have own power source and are inductively powered during communication with the reader. They are not capable to perform strong crypto operations (even symmetric encryption). Reader (*transceiver*) is a device with quite
big computational and energetic capabilities. Readers communicate with the
tags via radio channel. The last part of RFID system is a database that stores
information related with tags. Usually reader communicating with tags, uses a
database system.

Unfortunately, RFID technology entails some privacy threats. One of them
is tracking. For example, if a person is carrying an RFID-tag with static ID
with no encryption or blinding, then tracking is easy [4]. In this case tracking

© The Author(s) 2016
T. Czachórski et al. (Eds.): ISCIS 2016, CCIS 659, pp. 241–249, 2016.
DOI: 10.1007/978-3-319-47217-1_26

is understood as a possibility of identifying the tag. Another problem is that authentication here does not help much, because it is generally used in order to prevent revealing tag's stored data [9]. Tag's ID is usually not "masked". Thus learning tag's ID is quite easily achievable and sufficient for tag tracking.

In this paper a method for tracking prevention is described. We propose that tags has a dynamic ID. For this purpose, a tag should have built-in random number generator. We assume that tag's ID can be modified, for instance after every tag activation. Then the tag generates new ID and sends it to the reader which saves it in the system database. Considered is a passive model of an adversary who eavesdrops all the traffic, but not all the time [10]. If the adversary misses several changes of tag's ID, it may be not possible to identify again targeted tag. History of all tags IDs is stored in the backend database.

The rest of the paper is organized as follows: next section gives a short overview of methods for privacy preserving/tracking protection in RFID systems. Section 3 presents proposed method for tracking prevention. In Sect. 4 preliminary experimental evaluation of proposed method is presented; finally the last section concludes this work and gives possible future directions.

2   Related Works

The risk associated with privacy has been recognized quite quickly [2]. Unfortunately, some RFID systems do not use any security mechanisms, so tags can be read by any reader, which is an obvious threat to privacy [12]. For instance, an ability to identify a tag, can deliver information about its owner. It is then possible to create a profil of an user, based on information collected from tags [7]. Thus so far many techniques for privacy protection have been proposed. In [9], there is proposed a method for tracking prevention. Considered is a model, where an attacker monitors a large fraction of interactions, but not all of them. Authors propose to make small changes with the tag's identifier. Tag does not have to perform any cryptographic functions.

Another method is "masking" tags, described in [4,14]. It assumes that a tag stores a list of pseudonyms $p_1, p_2, \ldots, p_k$ and every now and then changes them. An adversary would not know that for example p_i and p_j belong to the same tag, therefore such approach can effectively complicate recognizing a tag. However, if an adversary intercepts tag's list of pseudonyms, the whole idea is compromised. Another question worth considering is how many pseudonyms should have store. Should be taken into account that tag has strongly limited memory resources [4].

Popular method is the *kill* command which aim is to completely deactivate a tag [12]. However this approach strongly reduces functionality of the system [8]. Another possible solutions are: screening with Faraday Cage or physical destruction of antenna or other parts of a tag [8]. More advanced solution is called *active jamming*. It is based on actively broadcasting radio signals, what disrupts actions of any reader. However, this approach requires extra device [11].

In [6] there is proposed an extension of method from [15], where tag can be temporarily switched off and another tag is simulating tags of all possible IDs. Hence a reader is not able to determine a tag which established a connection.

Golle et al. proposed in [5] a method called *universal re-encryption*. This solution is based on the classical scheme ElGamal which allows for re-encryption of a ciphertext without knowledge about public key. Thereby computationally powerful devices can read from a tag its content, then re-encrypt it and save it back in the tag. In this case only tag's owner, who knows the proper private key, is able to track the tag. Further development of this idea was proposed in [1].

3   A Method for Tracking Prevention

3.1   System and Privacy Model

We assume that RFID system consists of several tags, a reader and the backend database. More formal definition is presented in Definition 1.

Definition 1 (RFID system). *Let S denote RFID system. S consists of reader $\mathcal{R}$, finite set of i tags (transponders) $\mathcal{T} = \{T_1, T_2, \ldots, T_i\}$ and database $\mathcal{DB}$ which stores information related with the tags. $\mathcal{DB}$ also stores for each tag $\mathcal{ID} = \{ID_1, ID_2, \ldots ID_n\}$ which is the history of all tags' IDs. ID_n is defined as history of IDs of tag's n: $ID_n = \{ID_n^1, ID_n^2, \ldots, ID_n^k\}$, where ID_n^k is the k-th ID of the n-th tag.*

It is assumed that tags are passive (powered only during the communication with the reader).

In Definition 2 we introduce a simple model of an adversary and his goals. We define adversary's goal similarly as in the scheme proposed in [3]. A passive adversary $\mathcal{A}$ eavesdrops all the communication between RFID system components (i.e. the *forward* and *backward channel*), but not all the time.

Definition 2 (Adversary's goal – *unlinkability* game). *Suppose that there exists list of n tags IDs: $\mathcal{ID} = \{ID_1, ID_2, \ldots ID_n\}$, where ID_n is defined as in Definition 1. Then, it is choosed $ID_x^k \in \mathcal{ID}$ which is the currently used ID of some tag $T_x \in \mathcal{T}$. The goal of the adversary is to guess x with the probability greater than $\frac{1}{n}$.*

In our approach we assume that adversary observing the communication between reader and a tag, can "miss" several queries. The goal of the adversary is to identify the tag, i.e. not to "lose" its ID.

3.2   Tracking Prevention

We propose a method ChangeID which can be used to make more difficult recognition a particular tag. This method assumes that a tag simply changes its own identifier by generating a new one. Then, a new ID is transferred to the reader which saves it in the backend database. This makes possible later identifying the tag. Below is presented an idea of method ChangeID.

1. Tag has a n-bit binary sequence which stands for its ID: $(b_1, \ldots, b_n) \in \{0, 1\}^n$;
2. Next n bits are overwritten at random: a new sequence is created $(b_{i_1}, \ldots, b_{i_n})$, where for all $j \leq n$, $b_{i_j} \leftarrow b \in_U \{0, 1\}$ is substituted from a uniform distribution.

This procedure can be performed after each activation of tag or, for instance at specified intervals. Note that none of sensitive data is transferred through the forward channel which is assumed to be easily eavesdropped [11, 15]. It is likely that at average $n/2$ bits could remain unchanged.

Formally, this approach can be described as Algorithm 1.

ChangeID
Input: $(b_1, \ldots, b_n) \in \{0, 1\}^n$
Output: $(b_{i_1}, \ldots, b_{i_n})$

for $j \leq n$ **do**
 | $b_{i_j} \leftarrow b \in_U \{0, 1\}$
end

Algorithm 1: ChangeID procedure

Note that this procedure has low requirements in terms of computational complexity.

3.3   Problem of Ambiguity

One should consider that generating random IDs may cause generation of two (or more) the same IDs. Such a situation is undesirable in most systems and sometimes can be critical to their functioning. Although intuitively the probability of happening such situation is quite small, one can assume that the reader (after each changing tag's ID) checks in the backend database, if generated ID already exists. If does, then tag simply could be asked to perform another ChangeID operation. Similarly, if new generated ID is the same as the previous one, another performance of ChangeID could be done. In this case we assume that considered is a sequential access model. This situation is presented in Table 1.

4   Preliminary Experimental Evaluation

We conducted a simple experiment in which we implemented a function generating different lengths random sequences (strings) that could act as a tag identifier. We checked the possible links between distances of these sequences and examined Hamming distances between them.

Table 1. ChangeID protocol

Reader	Tag

$\xrightarrow{\text{hello}}$

$s =$ ChangeID

$\xleftarrow{s}$

if s exists in $\mathcal{DB}$ **then**
query for another ChangeID
else save s

We divided an experiment into 5 trials, in each trial 80 sequences of the following lengths were generated:

1. 32 bits length;
2. 64 bits length;
3. 128 bits length;
4. 256 bits length;
5. 512 bits length.

We analyzed Hamming distances between sequences in each trial (for example, sequence (1) with sequence (2); (2) with (3), ...). For the clarity, we normalized results of Hamming distance on the interval $[0, 1]$.

4.1 Distances in 32 Bits Trial

In Fig. 1 there are presented distances between adjacent sequences in 32-bits trial. Similarity is mostly at the level 0.7–0.9.

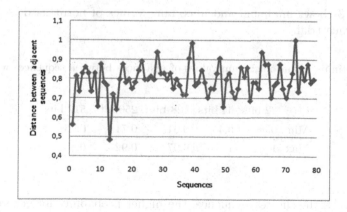

Fig. 1. Distances between adjacent sequences (total number of sequences: 80)

On the X-axis the are next sequences; Y-axis presents the normalized distance between adjacent sequences.

Table 2. Fragment of generated sequences for 32 bits trial

	Generated sequence	H_d	Norm
(1)	11011111001011011001010011001010		
(2)	10111100010110011100111100110100	21	0.66
(3)	10100010100111001110100010010111	27	0.84
(4)	11010100000100011111101000001101	18	0.56
(5)	11111010000100001100111000001010	21	0.66
(6)	11111110000100111000010011000011	22	0.69
(7)	10111010000010001011000000100011	28	0.88
(8)	11001000101011011101011110100000	25	0.78
(9)	11000010000100010110110110101111	27	0.84
...	...		
(79)	10111010001000111000110001111010		
(80)	11100010101001011101101100111000	23	0.72

In Table 2 there are presented several generated sequences and distances between adjacent sequences. H_d for i-th sequence stands for Hamming distance between the $i-1$ and i sequence, $Norm$ denotes value of normalization at $[0,1]$. For instance, H_d between (1) and (2) equals 21; in normalized way: 0.66, and so on.

For the clarity, we do not present full results of this and the other trials.

4.2   Summary

The Table 3 shows minimum and maximum values of normalized at $[0,1]$ distances in each trial.

Table 3. Minimum and maximum values of distances between sequences within each trial

	32 bits	64 bits	128 bits	256 bits	512 bits
Min	0.38	0.48	0.61	0.71	0.76
Max	1	1	0.97	0.92	0.88

Intuitively, the shortest sequence, the higher probability for generating two quite similar sequences (minimum distance for 32 bits is 0.38, for 64 bits – 0.48). The longer sequence, the greater differences (for instance, 0.76 for 512 bits sequences). These results are also showed in Figs. 2 and 3, respectively.

The longer tag's ID, the smaller probability of generating two the same sequences; however longer sequence requires more tag's memory.

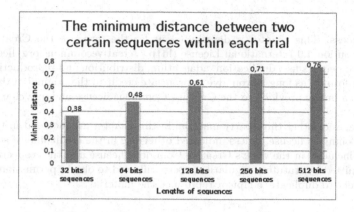

Fig. 2. The minimum (normalized) Hamming distance within each trials

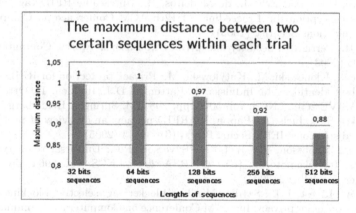

Fig. 3. The maximum normalized Hamming distance within each trials

5   Conclusion and Future Works

In this paper, a method for tracking prevention for RFID-tags was proposed. It was assumed that tag is able to change its own identifier by generating a random sequence and replacing earlier ID. If an adversary is not able to monitor the tag all the time, this method after a certain amount of execution can effectively complicate recognition of the tag. Preliminary experimental evaluation showed that unlinkability between tags IDs is at satisfactory level.

If future works it is planned to give a formal estimation of minimal number of ID modification in order to achieve good level of privacy. Also a simulation of implementation is considered to be carried out. Another problem to consider is to propose a method for settlement of the ambiguity of tags' IDs not in the sequential access model but in situation of independent and parallel operations of (several) readers.

References

1. Ateniese, G., Camenisch, J., de Medeiros, B.: Untracable RFID tags via insubvertible encryption. In: Proceedings of 12th ACM Conference on Computer and Communications Security (2005)
2. Chan, H., Perrig, A.: Security and privacy in sensor networks. Computer **36**(10), 103–105 (2003)
3. Cichoń, J., Klonowski, M., Kutyłowski, M.: Privacy protection for RFID with hidden subset identifiers. In: Indulska, J., Patterson, D.J., Rodden, T., Ott, M. (eds.) PERVASIVE 2008. LNCS, vol. 5013, pp. 298–314. Springer, Heidelberg (2008)
4. Garfinkel, S.L., Juels, A., Pappu, R.: RFID privacy: an overview of problems and proposed solutions. IEEE Secur. Priv. **3**(3), 34–43 (2005)
5. Golle, P., Jakobsson, M., Juels, A., Syverson, P.F.: Universal re-encryption for mixnets. In: Okamoto, T. (ed.) CT-RSA 2004. LNCS, vol. 2964, pp. 163–178. Springer, Heidelberg (2004)
6. Juels, A., Rivest, R.L., Szydlo, M.: The blocker tag: selective blocking of RFID tags for consumer privacy. In: ACM Conference on Computer and Communications Security, pp. 103–111 (2003)
7. Karthikeyan, S., Nesterenko, M.: RFID security without extensive cryptography. In: Proceedings of the 3rd ACM Workshop on Security of Ad Hoc and Sensor Networks, pp. 63–67. ACM. New York (2005)
8. Klonowski, M.: Algorytmy zapewniajace anonimowość i ich matematyczna analiza. PhD Dissertation (in Polish), Wrocław University of Technology, Poland (2009)
9. Klonowski, M., Kutyłowski, M., Syga, P.: Chameleon RFID and tracking prevention. In: Radio Frequency Identification System Security, RFIDSec Asia 2013, pp. 17–29 (2013)
10. Kutyłowski, M.: Anonymity and rapid mixing in cryptographic protocols. In: The 4th Central European Conference on Cryptology, Wartacrypt (2004). http://kutylowski.im.pwr.wroc.pl/articles/warta2004.pdf. Accessed 13 Feb 2016
11. Luo, Z., Chan, T., Li, J.S.: A lightweight mutual authentication protocol for RFID networks. In: Proceedings of 2005 IEEE International Conference on e-Business Engineering (ICEBE 2005), IEEE Xplore, pp. 620–625 (2005)
12. Medaglia, C.M., Serbanati, A.: An overview of privacy and security issues in the Internet of Things. In: Giusto, D., Iera, A., Morabito, G., Atzori, L. (eds.) The Internet of Things: 20th Tyrrhenian Workshop on Digital Communications, pp. 389–395. Springer, New York (2010)

13. Singh, D., Tripathi, G., Jara, A.J.: A survey of Internet-of-Things: future vision, architecture, challenges and services. In: 2014 IEEE World Forum on Internet of Things (WF-IoT), pp. 287–292 (2014)
14. Vajda, I., Buttyan, L.: Lightweight authentication protocols for low-cost RFID tags. In: Laboratory of Cryptography and Systems Security (CrySyS) (2003)
15. Weis, S.A., Sarma, S.E., Rivest, R.L., Engels, D.W.: Security and privacy aspects of low-cost radio frequency identification systems. In: Hutter, D., Müller, G., Stephan, W., Ullmann, M. (eds.) Security in Pervasive Computing. LNCS, vol. 2802, pp. 201–212. Springer, Heidelberg (2004)

Image Processing and Computer Vision

Diagnosis of Degenerative Intervertebral Disc Disease with Deep Networks and SVM

Ayse Betul Oktay[1(✉)] and Yusuf Sinan Akgul[2]

[1] Department of Computer Engineering, Istanbul Medeniyet University,
34700 Istanbul, Turkey
abetul.oktay@medeniyet.edu.tr
[2] GTU Vision Lab, Gebze Technical University, Gebze, Kocaeli, Turkey
akgul@gtu.edu.tr

Abstract. Computer aided diagnosis of degenerative intervertebral disc disease is a challenging task which has been targeted many times by computer vision and image processing community. This paper proposes a deep network approach for the diagnosis of degenerative intervertebral disc disease. Different from the classical deep networks, our system uses non-linear filters between the network layers that introduce domain dependent information into the network training for a faster training with lesser amount of data. The proposed system takes advantage of the unsupervised feature extraction with deep networks while requiring only a small amount of training data, which is a major problem for medical image analysis where obtaining large amounts of patient data is very difficult. The method is validated on a dataset containing 102 lumbar MR images. State-of-the-art hand-crafted feature extraction algorithms are compared with the unsupervisedly learned features and the proposed method outperforms the hand-crafted features.

Keywords: Degenerative disc disease · Auto encoders · Deep network

1 Introduction

Low Back Pain (LBP) is the most common pain type with 27% and it is the leading cause of activity limitation in USA under the age of 45 [7]. LBP is strongly associated with degenerative disc disease (DDD) [6]. Computer Aided Diagnosis (CAD) of DDD from MR images (Fig. 1) is crucial for many reasons. First, the inter-variability and intra-variability between the radiologists are high [12] and these variabilities affect diagnosis and treatment processes. A CAD system may reduce these variabilities. Second, the computer-based evaluation of an MRI sequence would help the radiologists in decreasing the costs and speeding up the evaluation process. In the literature, many machine learning based approaches with hand-crafted features have been proposed for CAD of various intervertebral disc diseases from MR images [1,4,5,9].

In recent years, deep networks have been widely used in many fields and they produce state-of-the-art results [3,10]. However, deep learning of medical

© The Author(s) 2016
T. Czachórski et al. (Eds.): ISCIS 2016, CCIS 659, pp. 253–261, 2016.
DOI: 10.1007/978-3-319-47217-1_27

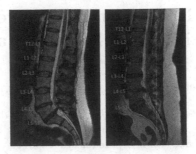

Fig. 1. Two MRI images that include the lumber region. The disc labels are shown on the images. The left image shows the discs L4-L5 and L5-S1. In the right image L3-L4 and L4-L5 discs are diagnosed as having DIDD

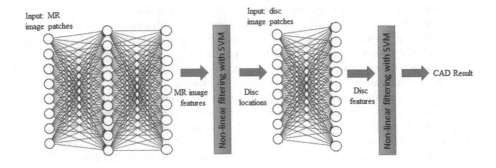

Fig. 2. The architecture of the system.

images has some domain-specific challenges. First, scaling the deep network for high dimensional medical images is mostly computationally intractable because of the large number of hidden neurons, often resulting in millions of parameters. Medical images have generally high resolution and the training needs high number of nodes. In addition, the large-scale data for training (even unlabeled) is not always available especially for many medical tasks where it is hard to gather data because of ethical issues. Furthermore, training data should involve many samples for different cases for CAD applications.

In this paper, we propose a novel deep learning architecture (Fig. 2) with non-linear filters that eliminates the requirement of large numbers of training data, network layers, and nodes. Instead of learning disc features with a traditional deep learning architecture, we propose to use non-linear filters together with auto-encoders [11]. The irrelevant input data is filtered with non-linear filters via SVM and only relevant data is fed to the succeeding layers. In this way, we restrict the upper layer to learn only the data that we consider valuable, which is very useful in reducing the training data size. Therefore, while the disc representations are learned with auto-encoders from the MR image patches, the non-linear filters reduce the domain of interest. Thus, with the first level non-linear filters the system focus on the discs from the whole MR image where the second level non-linear filters consider the disc representations for the diagnosis of DDD.

The method is tested and validated on a dataset containing 102 MR images. We also implemented the state-of-the-art features used in the methods of [1,2,9] and compared them with the features learned with auto encoders.

2   Unsupervised Feature Learning with Auto-encoders

An auto-encoder is a symmetrical neural network that aims to minimize the reconstruction error between the input and output data to learn the features. Let $X = \{x_1, x_2, ..., x_m\}$ be the image input for a single hidden layered auto-encoder where m is the input size. The output nodes are the same as the input nodes, thus the auto-encoder learns a nonlinear approximation of the identity function for estimating the output $\hat{X} = \{\hat{x_1}, \hat{x_2}, ..., \hat{x_m}\}$. Let k be the size of the nodes in the hidden layer and $W^{(1)} = \{w_{11}^{(1)}, w_{12}^{(1)}, ..., w_{km}^{(1)}\}$ be the weights where $w_{km}^{(1)}$ is the weight between input node m to hidden node k at hidden layer 1. The value of a hidden layer node is calculated by

$$z_i = b_i^{(1)} + \sum_{j=1}^{m} w_{ij}^{(1)} x_j, \tag{1}$$

where $b_i^{(1)}$ is the bias term for the node i at hidden layer 1. Each hidden node outputs a nonlinear activation function $a = f(z_i)$. The output layer $\hat{X}$ is constructed using the activations a as input and decoding bias and weights similar to Eq. 1. Features are learned by minimizing the reconstruction error of the likelihood function between X and $\hat{X}$ and the features are encapsulated in weights W. Backpropagation via gradient descent algorithm is used for adjusting W. Stacked auto-encoders are formed by stacking auto encoders by wiring the learned weights to the next auto encoder's input.

2.1   Intervertebral Disc Detection

In the proposed architecture, first the lumbar MRI features are learned with stacked auto-encoders. Let $d = \{d_1, d_2, ..., d_6\}$ be the labels of the lumbar intervertebral discs in an MR image. Our goal is to identify the location $l_i \in \Re^2$ of each disc d_i on the image I. Randomly selected patches from image I are used for learning the features of the images. Let β be a patch of size $m \times n$ of image I where m and n varies between the minimum and maximum disc width and height in the training set, respectively. The image patch β is resized to $r \times r$ pixels and is formed into a $1 \times r^2$ vector to be used as an input of an autoencoder. Figure 3 shows the unsupervised learning of lumbar MR image features with an auto-encoder.

The stacked auto-encoder with $X = r^2$ input nodes is trained with the vectorized image patches β. The weights W of the final hidden layer are brought to

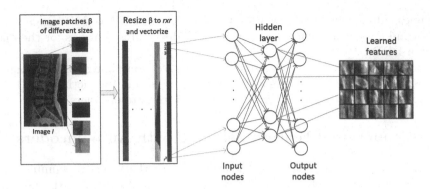

Fig. 3. An auto-encoder for learning MR image features. A single hidden layer auto-encoder trained with the vectorized image patches

square form (having $r \times r$ size) for building the feature set f of the MR images extracted in an unsupervised manner as explained in Sect. 2.

The feature set f includes the features of the whole MR image; however the objective of the proposed system is diagnosing the diseases related with the discs. To filter the irrelevant medical structures that exist in the image, we use nonlinear filtering with SVM. A sliding window approach is employed and each window $\Psi(p)$ enclosing the pixel p is convolved with the filter $f_i \in f$. The outputs of the convolution of each window with the filters in f are concatenated and the final feature vector is built. Each pixel p in the image I is given a score S_p with SVM that indicates the probability of being a location of disc d_i using f.

In order to locate and label the intervertebral lumbar discs, we follow the graphical model based labeling approach presented in [8] by enhancing the model with the unsupervised feature learning. We use a chain-like graphical model G consists of 6 nodes and 5 edges connecting the nodes where each lumbar intervertebral disc d_i is represented with a node. Our goal is to infer the optimal disc positions $d^* = \{d_1^*, d_2^*, ..., d_6^*\}$ where $d_i^* \in \Re^2$ and $1 \leq i \leq 6$ in the image I according to the given scores S_p and the spatial information between the discs in the training set. The optimal locations d^* of the discs are determined by using the maximum a posteriori estimate

$$d^* = \arg\max_d P(d|I, S_p, \alpha), \tag{2}$$

where I represents the image, S_p is the given score and α represents the parameters learned from the training set. The Gibbs distribution of $P(d|I, S_p, \alpha)$ is

$$P(d|I, P_s, \alpha) = \frac{1}{Z} exp \left\{ - \left[\sum \psi_L(I, d_k) + \lambda \sum \psi_{spa}(d_k, d_{k+1}, \alpha) \right] \right\}. \tag{3}$$

The function $\psi_L(I, d_k)$ represents the scores S_p given via deep learning and the potential energy function $\psi_{spa}(d_k, d_{k+1}, \alpha)$ captures the geometrical information between the neighboring discs d_k and d_{k+1}. The optimal solution d^* is gathered with dynamic programming in polynomial time. For the details of the graphical model G and inference, please refer to [8].

2.2 Diagnosis of DDD

After localizing the discs in the MR images, the disc features should be learned and they should be classified as healthy or not. The location l_i of each disc d_i is found with the Eq. 2. Since the window $\psi(p)$ enclosing the pixel p is known, these windows are directly used for CAD of degenerative disc disease. The windows $\Psi(p)$ of each located disc are used for training a sparse auto-encoder. The windows $\psi(p)$ are resized and vectorized to be used as input. The features are learned with sparse auto-encoders. The weights W of the final hidden layer of the auto-encoder are the used as the features f_d.

After determining the features of the discs, we again convolve the window $\psi(p)$ with the learned filter f_d. The output of the convolution operations are concatenated and the final feature vector is formed. These final feature vectors are trained and tested with SVM. Binary classification is performed and each window ψ is classified as having degenerative disc disease or not.

3 Experiments

In order to evaluate the proposed system, two different datasets, one with labeled and another with unlabeled discs, are used. First clinical MR image dataset contains the lumbar MR images of 102 subjects. The MR images are 512 × 512 pixels in size. In the images, there are 612 (102 subjects*6 discs) lumbar intervertebral discs where 349 of them are normal and 263 of them are diagnosed with degenerative disc disease. The disc boundaries are delineated and each disc is diagnosed having DDD or not by an experienced radiologist to be used as the ground truth. The second dataset includes the lumbar MR images of 43 subjects where the intervertebral discs are neither delineated nor diagnosed by an expert. This unlabeled dataset is used for providing data to the auto-encoder for unsupervised training. It is not used for testing the system since it does not include the ground truth.

For labeling process, randomly selected patches are used from the MR images. The width and height of the intervertebral discs are between 30–34 mm and 8–13 mm, respectively [13]. The patch size is selected in accordance with the intervertebral disc size. The total number of patches used for training is 10000. For preprocessing, the mean intensity value of the patch is subtracted from the image patch for normalization. The patches are resized to 15 × 15 pixels ($r = 15$) and the number of the input nodes X is 225. Two layers are used for the stacked auto encoder. The number of nodes in layer the first inner layer is 70 and the number of nodes in the second layer is 30.

The number of features f learned from the MR image patches is 30. Six-fold-cross-validation is used for SVR training. The parameters of the Eq. 3 are learned from the training set and the weighting parameter λ is selected as 0.5 empirically.

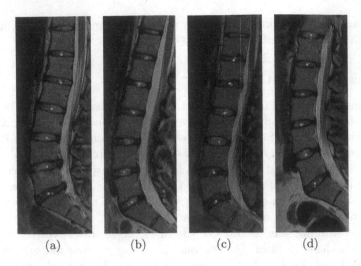

(a)   (b)   (c)   (d)

Fig. 4. Labeling results of the lumbar MR images selected from the database. Green rectangles are the ground truth center points and the red rectangles are the disc centers determined by our system. The MR images are cropped for better visualization (Color figure online)

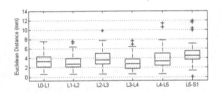

Fig. 5. Boxplot of the Euclidean distances of the disc centers determined by our system to the ground truth centers

Some of the visual labeling results of our system is shown in Fig. 4. In order to evaluate the performance of the labeling system with unsupervised feature learning, the Euclidean distances between the disc center point detected by our system and the ground truth are calculated. Figure 5 shows the boxplot of the Euclidean distances in mm.

For automated DDD diagnosis, a similar validation method is followed. Since the disc labels d determined for an image I and their enclosing windows ψ are determined in the labeling step, they are employed as the image patches for training and testing. Leave-one-out approach is used for training. Instead of using the whole window ψ, we use the half right side of the window ψ since the DDD including disc bulging and herniation occur at the right side. A two-layer stacked auto-encoder (70 nodes in the first layer, 40 nodes in the second layer) is employed for learning the features. The half right side of the labeled disc images are resized to 15×15 pixels in size and they are the input of the auto-encoder after vectorization.

Table 1. The accuracy, specificity, and sensitivity of the hand-crafted feature extraction methods and our method

Feature type	Number of features	Accuracy	Sensitivity	Specificity
Raw image intensity	1000	0.86	0.88	0.84
LBP	8	0.70	0.80	0.57
Gabor	1000	0.60	0.80	0.33
GLCM	5	0.71	0.78	0.62
Planar shape	3	0.55	1.0	0
Hu's moments	7	0.72	0.72	0.71
Intensity difference	12	0.89	**0.96**	0.82
Our method	40	**0.92**	0.94	**0.90**

After determining the features, each disc image is convolved with the features and the final feature vector for the final classification with binary SVM is created. The classification accuracy of the proposed system is 92 %.

In order to compare the unsupervised learned features with the hand-crafted features, popular feature types used in [1,9] are also implemented. The training is performed with six-fold-cross correlation and classification is performed via SVM. The number of features extracted and their accuracy, sensitivity, and specificity are reported in Table 1. The numerical results show that unsupervised learned features outperform hand-crafted features. The highest accuracy of the hand-crafted features 89.54 % for the intensity difference feature that calculates the numerical values (mean, standard deviation, etc.) of the intensities difference between T1-weighted and T2-weighted images. The accuracy of the unsupervised feature learning is higher than other hand-crafted features. In addition, the sensitivity and the specificity rates of the proposed system are higher than other state-of-the-art methods.

The experiments performed show that the DDD can be automatically diagnosed with a high accuracy with a few filters learned by auto-encoders. The unsupervised filters outperform other popular hand-crafted features even their number is lower than the hand-crafted features. In addition, the proposed system does not require a deep network structure including many hidden layers. The disc filters are efficiently learned with a two-layer auto-encoder with small training data.

4 Conclusions

In this paper, we present a novel method for CAD of the DDD with auto-encoders. The proposed architecture involves stacked auto-encoders and non-linear filters together for locating the intervertebral discs and diagnosis. The auto-encoders learns the image features effectively while the non-linear filters eliminates the irrelevant information. The system is validated on a real dataset of 102 subjects. The results showed that unsupervised learning of features yields a better representation and the features could be extracted with minimal user intervention. The comparison with popular hand-crafted features show that the results are comparable with the state of the art.

References

1. Ghosh, S., Alomari, R.S., Chaudhary, V., Dhillon, G.: Composite features for automatic diagnosis of intervertebral disc herniation from lumbar MRI. In: Conference of the IEEE Engineering in Medicine and Biology Society, pp. 5068–5071 (2011)
2. Ghosh, S., Alomari, R.S., Chaudhary, V., Dhillon, G.: Computer-aided diagnosis for lumbar MRI using heterogeneous classifiers. In: IEEE International Symposium on Biomedical Imaging, pp. 1179–1182 (2011)
3. He, K., Zhang, X., Ren, S., Sun, J.: Deep residual learning for image recognition. In: IEEE Conference on Computer Vision and Pattern Recognition (CVPR) (2016)
4. Koh, J., Chaudhary, V., Dhillon, G.: Diagnosis of disc herniation based on classifiers and features generated from spine MR images (2010)
5. Lootus, M., Kadir, T., Zisserman, A.: Radiological grading of spinal MRI. In: MICCAI Workshop: Computational Methods and Clinical Applications for Spine Imaging (2014)
6. Luoma, K., Riihimaumlki, H., Luukkonen, R., Raininko, R., Viikari-Juntura, E., Lamminen, A.: Low back pain in relation to lumbar disc degeneration. Spine 25(4), 487–492 (2000)
7. National Centers for Health Statistics: Chartbook on trends in the health of Americans, special feature: pain (2011). http://www.cdc.gov/nchs/data/hus/hus06.pdf/
8. Oktay, A.B., Akgul, Y.S.: Simultaneous localization of lumbar vertebrae and intervertebral discs with SVM-based MRF. IEEE Trans. Biomed. Eng. 60(9), 2375–2383 (2013)
9. Oktay, A.B., Albayrak, N.B., Akgul, Y.S.: Computer aided diagnosis of degenerative intervertebral disc diseases from lumbar MR images. Comput. Med. Imaging Graph. 38(7), 613–619 (2014)
10. Schroff, F., Kalenichenko, D., Philbin, J.: Facenet: a unified embedding for face recognition and clustering. In: The IEEE Conference on Computer Vision and Pattern Recognition (CVPR), June 2015
11. Tang, Y.: Deep learning using support vector machines (2013)

12. Van Rijn, J.C., Klemetsouml, N., Reitsma, J.B., Majoie, C.B.L.M., Hulsmans, F.J., Peul, W.C., Stam, J., Bossuyt, P.M., den Heeten, G.J.: Observer variation in MRI evaluation of patients suspected of lumbar disk herniation. AJR Am. J. Roentgenol. **184**(1), 299–303 (2005)

13. Zhou, S., McCarthy, I., McGregor, A., Coombs, R., Hughes, S.: Geometrical dimensions of the lower lumbar vertebrae - analysis of data from digitised CT images. Eur. Spine J. **9**(3), 242–248 (2000)

Output Domain Downscaler

Mert Büyükmıhçı[1], Vecdi Emre Levent[2], Aydin Emre Guzel[2], Ozgur Ates[2],
Mustafa Tosun[2], Toygar Akgün[3], Cengiz Erbas[3], Sezer Gören[1],
and Hasan Fatih Ugurdag[2(✉)]

[1] Department of Computer Engineering, Yeditepe University, Istanbul, Turkey
sgoren@cse.yeditepe.edu.tr
[2] Department of Electronics and Electrical Engineering,
Ozyegin University, Istanbul, Turkey
fatih.ugurdag@ozyegin.edu.tr
[3] ASELSAN, Ankara, Turkey
takgun@aselsan.com.tr

Abstract. This paper offers an area-efficient video downscaler hardware architecture, which we call Output Domain Downscaler (ODD). ODD is demonstrated through an implementation of the bilinear interpolation method combined with Edge Detection and Sharpening Spatial Filter. We compare ODD to a straight-forward implementation of the same combination of methods, which we call Input Domain Downscaler (IDD). IDD tries to output a new pixel of the downscaled video frame every time a new pixel of the original video frame is received. However, every once in a while, there is no downscaled pixel to produce, and hence, IDD stalls. IDD sometimes also skips a complete row of input pixels. ODD, on the other hand, spreads out the job of producing downscaled pixels almost uniformly over a frame. As a result, ODD is able to employ more resource sharing, i.e., can do the same job with fewer arithmetic units, thus offers a more area-efficient solution than IDD. In this paper, we explain how ODD and IDD work and also share their FPGA synthesis results.

1 Introduction

Downscalers are found in many image processing applications. This work addresses video streaming applications and hence needs to be real-time, which opens the door for hardware implementation.

Downscaling produces a lower resolution version of the input image. The purpose is to do this with the least quality loss in the image. The simplest downscaler in the literature is the Nearest Neighbor method (NN) [1]. NN is more area-efficient and easier to implement than other methods, for instance, Bicubic

This work has been partially supported by the Artemis JU Project ALMARVI (Algorithms, Design Methods, and Many Core Execution Platform for Low-Power Massive Data-Rate Video and Image Processing), Artemis GA 621439 [6] and TUBITAK (The Scientific and Technological Research Council of Turkey) Project number 114E343.

T. Czachórski et al. (Eds.): ISCIS 2016, CCIS 659, pp. 262–269, 2016.
DOI: 10.1007/978-3-319-47217-1_28

Interpolation (BcubI) [2] and Adaptable K-Nearest [3] methods. However, the drawback of NN is that the resulting image/frame contains blocking and aliasing artifacts. On the other hand, BcubI can handle blocking and aliasing issues well and produce high quality images; however, because of its complexity and memory requirements, its implementation is difficult and costly. A compromise is possible though. Another method, called Bilinear Interpolation (BlinI) [4], that can also handle blocking and aliasing issues, has lower complexity and hence lower cost than BcubI. Although its output has lower quality than BcubI, the downscaled images it produces are acceptable. Chen [5] proposes an enhanced BlinI downscaler that uses an edge detection algorithm and Sharpening Spatial Filter (SSF) before BlinI to prevent the blurring caused by BlinI.

In this paper, we propose a novel area-efficient implementation of the enhanced downscaler in [5]. We call our downscaler implementation Output Domain Downscaler (ODD) and the straight-forward implementation in [5] as Input Domain Downscaler (IDD). Note that both ODD and IDD apply to also other downscaling algorithms.

IDD tries to output a new pixel every time a new input pixel is received. However, once every few input pixels, there is no downscaled pixel to produce, and IDD stalls (i.e., idles). IDD sometimes also skips a complete row of input pixels. ODD, on the other hand, spreads out the job of producing downscaled pixels almost uniformly over a frame. As a result of that, ODD is able to do more resource sharing, i.e., can do the same job with fewer arithmetic units, thus offers a more area-efficient solution than IDD. In this paper, we implement our ODD architecture with a downscale ratio between 1 and 2 with no loss of generality. That is because it is best to achieve larger downscale ratios of BlinI by applying a downscale ratio between 1 and 2 multiple times. Note that we implemented Verilog RTL generators for ODD and IDD, which are highly parameterized, instead of implementing fixed instances of the two architectures with a specific downscale ratio, fps, and frame resolution. Besides datapath optimizations, we also did memory optimizations as well.

2 The Downscaling Algorithm

The downscaling algorithm implemented in this work is the algorithm in [5], which is based on BlinI. [5] proposes the idea of detecting edges and boosting the pixels around them with SSF in order to circumvent the blur caused by BlinI.

When Edge Detection (ED), SSF, and BlinI are considered altogether, a sliding of 8 input pixels shown in Fig. 1a are used around the downscaled pixel (e.g., pixels P, Q, R). These 8 pixels are used to decide the values of the 4 pixels (pointed to by the arrows) immediately around the downscaled pixel, which are then used by BlinI. In Fig. 1, the input pixels (the dots) are at integer locations, while the downscaled pixels of P, Q, R are at fractional locations with a distance of 1.5 between them, assuming that the downscale ratio is 1.5. If P is at an x coordinate of 1.3, then Q and R are at respectively 2.8 and 4.3. When we take the integer part of these coordinates, we get 1, 2, and 4. These numbers show

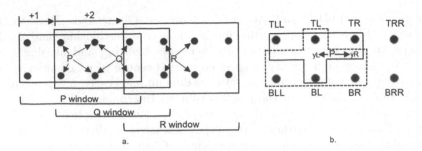

Fig. 1. a. ODD's sliding window b. SSF and BlinI's windows when edge is at L

the starting positions of these consecutive sliding windows. One way to describe this is that the sliding window sometimes shifts by 1 and sometimes by 2. This is our way of looking at it (i.e., the ODD way). Another way to look at this is that sliding window always shifts by 1 but sometimes it does not produce a downscaled pixel. This is the IDD way of looking at it.

Top 4 of these 8 pixels are used for ED. That are the pixels marked with TLL (Top Left Left), TL, TR, TRR as shown in Fig. 1b. In order to find if there are edges at pixel P, the Asymmetry parameter, A, for that pixel needs to be computed as defined by Eq. 1. If A is more positive than a positive threshold, it means that there is a vertical edge at the horizontal position of L (no horizontal edges are considered). If A is more negative than the negative of the same threshold, there is an edge at R.

$$A = |P_{TRR} - P_{TL}| - |P_{TR} - P_{TLL}| \qquad (1)$$

Suppose an edge is detected at the horizontal position of L (as opposed to R), then the T-like convolutional window in Fig. 1b is used to recompute the input pixel at location TL, which is the pixel where the edge is detected. The neighboring pixels are multiplied by -1 and pixel TL is multiplied by the sharpening coefficient, S, and the sum is divided by $S - 3$. The pixel below where the edge is detected (BL) is also recomputed by the SSF, hence the dotted window in Fig. 1b. If the edge is detected at R, then SSF shifts the two T-like windows to the right by one position. Hence, SSF uses all 8 pixels to compute two pixels and then replaces either TL and BL pixels or TR and BR.

BlinI computes a downscaled pixel as a weighted average of 4 input pixels surrounding it, i.e., TL, TR, BL, BR pixels. To compute output pixel P, which we also denote by P_{xy}, we first compute two intermediate pixel values (Eqs. 2 and 3, namely, P_{yL} and P_{yR} (see Fig. 1b for locations of yL and yR), as weighted averages of pixels vertically positioned with respect to them, where dy is the weight of the bottom pixel and $1 - dy$ is the weight of the top pixel. Then, we take a weighted average of the two intermediate pixels to compute the pixel value at downscale location (x, y) and arrive at Eq. 4. Note that dx and dy are respectively fractional parts the x and y coordinates of the downscaled pixel P, in other words, they constitute the displacement of P from input pixel TL.

$$P_{yL} = (P_{BL} - P_{TL})dy + P_{TL} \tag{2}$$
$$P_{yR} = (P_{BR} - P_{TR})dy + P_{TR} \tag{3}$$
$$P = P_{xy} = (P_{yR} - P_{yL})dx + P_{yL} \tag{4}$$

3   Output Domain Downscaler

Consider a video stream at 90 frames per second (fps) and full HD resolution (1920 by 1080 pixels per frame). If the downscaler is running at a clock frequency of 187 MHz, then we will be receiving one input pixel per clock cycle. If we designed the hardware of our downscaler in a brute-force manner (i.e., the IDD way), then we would be shifting our sliding window of 8 input pixels to the right by one pixel every clock cycle just like most designers do in most video streaming applications.

Consider a downscale ratio of 1.8. Then, we would be producing 1067 downscaled output pixels per one line of a video frame. That is, we would be idling in 853 (=1920 − 1067) non-consecutive cycles. We would also be idling for 360 complete lines, each time 1920 cycles back to back. That is because the step size in the vertical direction is also equal to the downscale ratio.

However, since sometimes we would need to produce downscaled pixels in back to back cycles, we would have to design an arithmetic datapath that can execute all operations at a throughput (but not necessarily latency) of 1 downscaled pixel per 1 cycle. Therefore, we would not be able to do resource sharing and would employ as many multipliers as multiplication operations, as many adders as addition operations, and so on.

Fortunately, we do not do it that way; we do it as follows. While IDD shifts the sliding window by one position every time a new input pixel is received (i.e., once every Input Cycle Time, or in short, ICT), we slide the window by the scale ratio, 1.8, in a time period of 3 times ICT (i.e., Output Cycle Time, or in short, OCT). If ICT is 1 cycles per input pixel, then our OCT is 3 cycles per output pixel.

OCT is 3 because we produce N/r^2 output pixels over one frame time if there are N pixels in an input frame. If $r = 1.8$, then we could spread our computations for a downscaled pixel over 3.24 cycles, it would be perfect. However, we have to schedule computations over an integer number of cycles unless we are willing to do loop unrolling. To summarize, $OCT = \lfloor ICT * r^2 \rfloor$.

In our ODD architecture, Output Cycle Time (OCT) determines the cycle time of the datapath (i.e., hence length of the schedule), and that is why it is called "Output Domain". On the other hand, in the naive IDD approach, Input Cycle Time (ICT) determines the cycle time of the datapath, hence the name "Input Domain". OCT is larger than or equal to ICT; therefore, ODD has more opportunity for resource sharing, and in the asymptotic case, uses M/r^2 arithmetic units, whereas IDD uses M arithmetic units.

Fig. 2. a. IDD's top-level b. ODD's top-level

Figure 2 shows the top levels of ODD and IDD architectures. Both ODD and IDD employ a line buffer (Linebuf) and a FIFO. ODD's datapath is connected to the output port of the FIFO, while IDD's datapath is on the input side of its FIFO. Line buffers are, on the other hand, 1 line and 4 pixel long and are due to the 4 × 2 sliding window the downscaling algorithm uses (shown in Fig. 1).

It is obvious that ODD needs a FIFO. While input pixels are received in raster order at a rate of 1 pixel per cycle, ODD consumes them at a rate of 1.8 pixels (due to the downscale ratio) every 3 cycles. Therefore, it consumes $1.8/3 = 0.6$ pixels per cycle, and as a result the FIFO of input pixels builds up at a rate of 0.4 pixels per cycle. When the downscaler skips a line, then it catches up. It even sometimes leapfrogs the input pixels and waits for the FIFO to fill up as it has a cycle-time of 3 cycles as opposed to the ideal and slower rate of 3.24 cycles.

On the other hand, it is not obvious that IDD needs a FIFO. However, if we have a non-stallable pipeline at the output of the downscaler, and/or we desire to minimize the amount of logic in that pipeline, we need to buffer the downscaled pixels in a FIFO and spread out the computations in the video pipeline that uses the downscaled frames over a pipeline heart-beat of $\lfloor ICT * r^2 \rfloor$ cycles.

ODD's FIFO is a special FIFO; unlike a regular FIFO, it has different width on the write and read sides. It is 1-pixel wide on the write side and 8-pixel wide on the read side. It is indeed a FIFO as all it needs is a push/pop interface with addresses (i.e., write and read pointers) kept inside. Its write pointer is the coordinates of the input pixel that is being received. Its read pointer is the coordinates of the downscaled pixel that is being currently worked on. However, the FIFO outputs 8 input pixels with addresses based on some arithmetic done with the fractional read pointer. Note that in ODD's case, Linebuf can be merged into the FIFO.

Figure 3a gives a procedural code for the downscaling algorithm implemented in this work. Figure 3b shows its Data Flow Graph (DFG). The schedule obtained by mapping this DFG to arithmetic units (columns of the schedule) is shown in Fig. 3c. Every operation in the DFG is named after its output variable. The subscripts of the variable (thus operation) names in the schedule indicate the index of the output pixel, i.e., its order in the video stream. We scheduled ED, SSF, and BlinI separately.

While [5] does all computations in fixed point arithmetic, we do BlinI part in floating point arithmetic since the algorithmic verification model we are given by our image processing people does BlinI in floating point. The advantage of floating point is that it eliminates the engineering time to fine tune the decimal point location in fixed point. Therefore, ED and SSF use integer arithmetic units

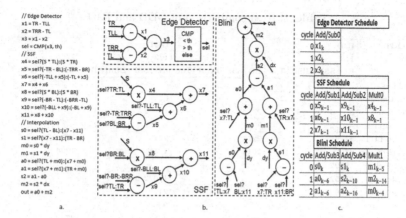

Fig. 3. a. Downscaling algorithm b. Its DFG c. Its schedule for OCT = 3

(non-pipelined), while BlinI uses heavily pipelined floating point units, which is why the degree of functional pipelining in BlinI is quite high ($k - (k - 14) + 1 = 15$ stages).

4 Synthesis Results

We implemented our architecture not as a fixed RTL design but as a Perl generator that outputs a Verilog RTL design, given design parameters of fps, resolution, clock frequency, and downscale ratio. We targeted a Virtex-7 FPGA. We obtained synthesis results for 90 fps, 1920×1080 pixels/frame, clock frequency of 187 MHz, and a downscale ratio of 1.8 for both ODD and IDD.

Hardware resources needed for both ODD and IDD are given in Table 1. Note that FP stands for Floating Point. FP Adders are in fact Add/Sub units. Int. stands for Integer. Although IDD does BlinI with 2 FP multiplications and 4 FP additions/subtractions as opposed to ODD's 3 and 6, respectively, ODD still uses substantially fewer hardware resources.

We have generated and synthesized ODD and IDD for two different cases. One case has an ICT of 1, and the other has an ICT of 2. When OCT is computed for the downscale ratio of 1.8 for these cases, we obtain 3 and 6. Therefore, we have ICT/OCT of 1/3 and 2/6 for these cases.

Linebuf is the same size for both ODD and IDD; however, the FIFO size is different. IDD has a FIFO that is more shallow but wider. That is because it sores the output pixels, which have a $1/1.8$ times the rate of input pixels and are wider (32 bits versus 8 bits). Hence, IDD FIFO is $4/1.8$ times (45 % of) ODD FIFO. When Linebuf is also taken into account, the memory part of ODD is approximately 60 % of IDD. These numbers are the same for both 1/3 and 2/6 cases.

As for the Datapath, Table 1 first lists the number of arithmetic units per subtask of the downscaler (ED, SSF, BlinI) and the total numbers (Tot.). The number of LUTs and flops these arithmetic units amount to are listed on the lines in

Table 1. Area comparison of ODD and IDD

ICT/OCT	IDD								ODD							
	1/3				2/6				1/3				2/6			
	ED	SSF	BlinI	Tot.	ED	SSF	BlinI	Tot.	ED	SSF	BlinI	Tot.	ED	SSF	BlinI	Tot.
FP Adders	–	–	4	4	–	–	2	2	–	–	2	2	–	–	1	1
FP Multipliers	–	–	2	2	–	–	1	1	–	–	2	2	–	–	1	1
Int. Adders	3	6	–	9	2	3	–	5	1	2	–	3	1	1	–	2
Int. Multipliers	–	2	–	2	–	1	–	1	–	2	–	2	–	1	–	1
Datapath LUTs	4499				2276				2215				1550			
Datapath Flops	3797				2012				1958				1294			
Linebuf Mem.	15392 bits															
FIFO Mem.	37952 bits								17072 bits							
Memory LUTs	3569								2172							
Memory Flops	182								98							
Total LUTs	8068				5845				4387				3722			
Total Flops	3979				2194				2056				1392			

Table 1 that start with "Datapath LUTs" and "Datapath Flops". The hardware resources ODD needs for the Datapath (LUTs and Flops) are roughly half of what IDD needs in 1/3 case, while it is two thirds in 2/6 case. When we look at the total needed (Datapath + Memory), in 1/3 case ODD requires 54 % of IDD in terms of LUTs and requires 52 % of IDD in terms of flops. Those numbers are 64 % and 63 %, respectively, for the 2/6 case.

5   Conclusion

In this paper, an area-efficient downscaler hardware architecture, called Output Domain Downscaler (ODD) was presented. ODD was compared to Input Domain Downscaler (IDD) architecture, which is the straight-forward approach used in pretty much all downscaler hardware implementations. While ODD is applicable to every downscale algorithm, we have implemented ODD for the downscale algorithm in [5] to show its merits. Our only modification is the use of floating point instead of fixed point in the interpolation stage. We have implemented the same algorithm with IDD as well. We produced ODD and IDD designs from our ODD and IDD Verilog RTL generators for two different cases of input/output rates. We found that ODD uses roughly half the hardware resources of IDD in one case and two thirds in the other case. Hence, we suggest ODD as a viable architecture for a variety of downscale algorithms.

References

1. Caselles, V., Morel, J.M., Sbert, C.: An axiomatic approach to image interpolation. IEEE Trans. Image Process. **7**(3), 376–386 (1998)
2. Nuno-Maganda, M.A., Arias-Estrada, M.O.: Real-time FPGA-based architecture for bicubic interpolation: an application for digital image scaling. In: International Conference on Reconfigurable Computing and FPGAs (ReConFig 2005), Puebla City, pp. 1–8 (2005)
3. Ni, K.S., Nguyen, T.Q.: Adaptable K-nearest neighbor for image interpolation. In: IEEE International Conference on Acoustics, Speech and Signal Processing, Las Vegas, pp. 1297–1300 (2008)
4. Jensen, K., Anastassiou, D.: Subpixel edge localization and the interpolation of still images. IEEE Trans. Image Process. **4**(3), 285–295 (1995)
5. Chen, S.L.: VLSI implementation of an adaptive edge-enhanced image scalar for real-time multimedia applications. IEEE Trans. Circuits Syst. Video Technol. **23**(9), 1510–1522 (2013)
6. Artemis JU Project ALMARVI Algorithms, Design Methods, and Many-CoreExecution Platform for Low-Power Massive Data-Rate Video and Image-Processing, GA 621439. http://www.almarvi.eu

The Modified Amplitude-Modulated Screening Technology for the High Printing Quality

Ivanna Dronjuk[1], Maria Nazarkevych[2(✉)], and Oksana Troyan[2]

[1] Automated Control Systems Department Institute of Computer Science,
National University Lviv Polytechnic, Lviv, Ukraine
idronjuk@polynet.lviv.ua

[2] Publishing Information Technology Department Institute of Computer Science,
National University Lviv Polytechnic, Lviv, Ukraine
mar.nazarkevych@gmail.com, troyan@gmail.com

Abstract. A new screening method based on the new form of screening element in improving printing quality was considered. The relationship between the Ateb-functions and the generalized superellipse is proved. Printing quality is an essential parameter when incorporating specially designed security features into the electronic file from which printing is done. Advisability of applying the proposed method for protection of information on the physical media was analyzed.

1 Introduction

The printing technology in computer epoch is completely changed. All details are described in classical books [1,2]. Digital screening is considered an algorithmic process that creates the images from an arrangement of small, binary dot elements. Generally in the different approaches for half-toning are two main screening methods: Amplitude Modulated and Frequency Modulated. Comparison of these two methods is described in [3]. Problem of improving printing quality using screening is concerned in [4]. The purpose of this study is to develop a modified amplitude-modulated screening method to improve the print quality. Improving the screening process can more accurately reflect the subtle elements of the image or text which makes protection of printed information on the physical media more reliable.

To implement the task, special protective graphics based on periodic Ateb-functions were built and the method of modified amplitude-modulated screening that allows the realization of printing fine detail and halftones with greater clarity was proposed.

This article continues the study, which was beginning in [5]. The modified amplitude-modulated screening technology allows to print small contours, lines and halftones with maximal precision.

2 Mathematical Model

Let us consider oscillation, as a nonlinear oscillating system with one degree of freedom. Modeling behavior of the system $x(t), y(t)$ is generated by a system of an ordinary differential equations in the form

© The Author(s) 2016
T. Czachórski et al. (Eds.): ISCIS 2016, CCIS 659, pp. 270–276, 2016.
DOI: 10.1007/978-3-319-47217-1_29

$$\begin{cases} \frac{dx}{dt} + \beta y^m = 0, \\ \frac{dy}{dt} + \alpha x^n = 0. \end{cases} \tag{1}$$

where $x(t), y(t)$ – are a values at time t; α, β – constants that determine size of the oscillation period; n, m – numbers that determine the degree of nonlinearity of the equation that affects the period of the main component of fluctuations. In the performance of such conditions on α, β and n, m : $\alpha \neq 0, \beta \neq 0, n = \frac{2k_1+1}{2k_2+1}, k_1, k_2 = 0, 1, 2 \ldots, m = \frac{2p_1+1}{2p_2+1}, p_1, p_2 = 0, 1, 2 \ldots$ it is proved [6], that the analytical solution of equation (1) is represented as Ateb - functions.

The solution (1) is represented through periodic Ateb-functions [6] as follows

$$\begin{cases} x = C_1 Ca(n, m, \phi), \\ y = C_2 Sa(m, n, \phi). \end{cases} \tag{2}$$

where C_1, C_2 are the some constants, $Ca(n, m, \phi), Sa(m, n, \phi)$ are Ateb-cosine and Ateb-sine respectively. Variable ϕ is associated with time t as follows

$$\phi = C_3 t + \phi_0, \tag{3}$$

where C_3 - is some constant, ϕ_0 - the initial phase of the oscillations, which are determined from the initial and periodical conditions for the system (1).

Periodical conditions are presented by expressions

$$\begin{cases} Ca(n, m, \phi + 2\Pi) = Ca(n, m, \phi), \\ Sa(m, n, \phi + 2\Pi) = Sa(m, n, \phi). \end{cases} \tag{4}$$

where Π is a half period of Ateb-function. Taking into account identity [2]

$$Ca(n, m, \phi)^{m+1} + Sa(m, n, \phi)^{n+1} = 1, \tag{5}$$

we result following formula for a half period of Ateb-functions

$$\Pi(m, n) = \frac{\Gamma(\frac{1}{n+1})\Gamma(\frac{1}{m+1})}{\Gamma(\frac{1}{n+1} + \frac{1}{m+1})}. \tag{6}$$

In formula (6) denomination $\Gamma(\bullet)$ means Gamma function. Identity (5) is a generalization of well-known trigonometrical identity $\cos^2\phi + \sin^2\phi = 1$ in the case of Ateb-functions. So Ateb-functions generalize trigonometrical functions, if parameters $n = 1$ and $m = 1$, than $Ca(1, 1, \phi) = cos\phi$ and $Sa(1, 1, \phi) = \sin\phi$.

3   A Relationship Between the Ateb-Functions and the Generalized Superellipse

In this section we show the relationship between the Ateb-functions and plane algebraic Lame's curves which is known also as a generalized superellipse. We propose to construct a unique raster element based on the Ateb-functions which we transform in a graphic element as a generalized superellipse. Representation superellipse by Ateb-functions enables a functional control under the screening element. Consider the generalized superellipse formula as follows [7]

$$\left|\frac{x}{A}\right|^p + \left|\frac{y}{B}\right|^q = 1;\ p, q > 0, \tag{7}$$

where p, q, A and B are positive numbers. Let we substitute (2) into formula (7) we have a new formula

$$\left|\frac{C_1 Ca(n, m, \phi)}{A}\right|^p + \left|\frac{C_2 Sa(m, n, \phi)}{B}\right|^q = 1. \tag{8}$$

If we define $p = m + 1, q = n + 1$ and will select A and B that satisfy conditions $\frac{C_1}{A} = 1, \frac{C_2}{B} = 1$, we obtained exactly identity (5). Identity (8) shows a relationship between the Ateb-functions and the generalized superellipse. Thus we prove a new fact that main Ateb-function identity can be presented as a the generalized superellipse formula and periodical Ateb-cosine and Ateb-sine are strongly connected to the generalized superellipse.

We use formula (8) under conditions $n = m$ corresponding to the superellipse (not generalized) for constructing a new screening element. If we define $A_1 = \frac{A}{C_1}$ and $B_1 = \frac{B}{C_2}$ we obtain the curve a new generalization of the generalized superellipse as

$$\left|\frac{Ca(n, m, \phi)}{A_1}\right|^p + \left|\frac{Sa(m, n, \phi)}{B_1}\right|^q = 1. \tag{9}$$

The further generalization of the superellipse is given in polar coordinates (r, ϕ) in case $r \neq 1$ by

$$r(n, m, \phi) = \left|\frac{Ca(n, m, \phi)}{A_1}\right|^p + \left|\frac{Sa(m, n, \phi)}{B_1}\right|^q. \tag{10}$$

We propose to name it the *Ateb-superellipse*. The area S inside the superellipse can be expressed in terms of the Gamma function as

$$S = 4^{1 - \frac{1}{n+1}} \sqrt{\pi} AB \frac{\Gamma(1 + \frac{1}{n+1})}{\Gamma(\frac{1}{2} + \frac{1}{n+1})}, \tag{11}$$

where S defines the area of the proposed screening element.

4   Technological Characteristics of the Screening Method

A secure document must comply with International Standard ISO 14298:2013 specifies requirements for a security printing management system for security printers [8]. Safety elements should be made within 40–50 microns positive play and 60–80 microns reversed, and microprint size should be within 200–250 microns, which guarantees high quality of printing and helps to reduce the likelihood of fraud. The authors have developed a new method for screening

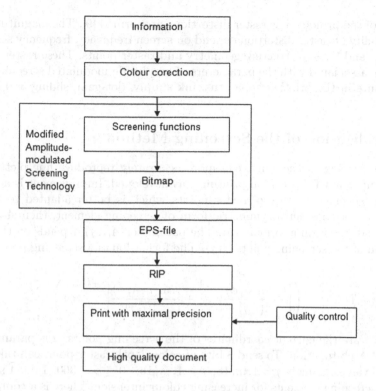

Fig. 1. Block diagram of screening technology

technology for improving printing quality. Figure 1 shows a block diagram of the proposed method. The resolution ability of print is restricted by the capacities of the output printing device.

It is important to provide high quality of the imprint for effective data protection on physical media. The better printed information is, the harder it is to forge it. Modern technologies allow faking everything, but there arises a question of economic criteria, namely the time and the cost of creating a fake. The main purpose of defense is to make the fake unprofitable. It is clear that the increase of print quality leads to higher cost of printed impression, and thus the cost of fraud rises. This is especially important for full-color prints, which are the most important documents (passport, driving license, etc.).

There is a problem of converting structure images in the process of printing, which is related to the difficulty of rendering fine detail and halftones. One of the most significant shortcomings of modern methods of structural transformation is much smaller resolution of the prints compared to the resolving ability of printing. This is due to the amplitude-modulated principle with binary halftone reproduction means of printing in which the tone values in a particular area of the original play with the relative area of the colored area of the print. Raster points are destroying contours and fine detail of halftone original, reducing the quality of the prints. Thus raster distortions are formed [9]. The magnitude and the visibility of raster distortion depend on screen frequency, frequency scanning function, and bitmap structure geometry and raster points. These raster distortions are associated with the parameters of amplitude-modulated screening such as pressure in the printing apparatus, ink supply, dot gain, sliding and double vision.

5   Realization of the Screening Method

A new screening method that can more accurately reproduce fine picture elements important for precision printing was developed. Improvement is achieved with the special structure of raster points which is better adapted to display halftones. Let consider a symmetric form of screening element, then $A = B = A(i, j)$ and $n = m$ in a formula (8). The parameter $A(i, j)$ depends on the color intension of the screening points (i, j). The formation of a screening point is the formula:

$$T(i,j) = \left(\left| \frac{Ca(n, n, \phi)}{A(i,j)} \right|^{n+1} + \left| \frac{Sa(n, n, \phi)}{A(i,j)} \right|^{n+1} = 1 \right)_{(i,j)} \quad (12)$$

where i, j are the current coordinates of the screening points, n is parameter of periodic Ateb-function. To send 8 bits of color depth raster point can take from 1 to 256 values, namely $j = 1, ..., 16$; $i = 1, ..., 16$; $0 \leq \phi \leq 360$. Table 1 shows a unique screening elements for increasing colour intension. There is a comparison of a standard circle (row 1) and proposed screening elements (row 2). Table 2 presents calculation of the unique screening elements with parameter $n + 1 = e$ for colour intention from 5 to 100 %. For screening element we represent the colour intention as an area S of screening element, where S is calculating with formula (11). The point with a darker colour has a bigger screening element.

Development of the modified method of autotypical screening allows printing the fine details and halftones for text or graphical information on a physical medium more precisely which is shown in a Fig. 2. Figure 2 shows a large scale result of the screening method. The halftone reproduction is better for an image (b) than an image (a) for a normal size.

Table 1. The comparison of a standard circle and the unique elements of screening technology

Form Screening Functions									
10 %	20 %	30%	40 %	50 %	60 %	70 %	80 %	90 %	100%
·	●	●	●	●	●	●	●	●	●
·	·	▪	▪	▪	▪	▪	▪	▪	▪

Table 2. Calculation table of the superellipse screening elements

Superellipse				
Basic parameters			Perimeter	Area
A full size width (mkm) float	B full size height (mkm) float	n + 1 = e Ateb-parameter, float	P, mkm	S, mkm^2
2.4	2.4	2.718	7.97	5.00
3.5	3.5	2.718	11.62	10.01
4.2	4.2	2.718	13.94	15.00
4.9	4.9	2.718	16.27	20.10
5.4	5.4	2.718	17.93	25.14
5.9	5.9	2.718	19.59	30.00
6.4	6.4	2.718	21.25	35.21
6.8	6.8	2.718	22.58	40.00
7.2	7.2	2.718	23.90	45.00
7.6	7.6	2.718	25.23	50.02
8	8	2.718	26.56	55.33
8.4	8.4	2.718	27.89	60.00
8.7	8.7	2.718	28.88	65.22
9	9	2.718	29.88	70.00
9.3	9.3	2.718	30.88	75.00
9.6	9.6	2.718	31.87	80.11
9.9	9.9	2.718	32.87	85.00
10.3	10.3	2.718	34.20	90.59
10.5	10.5	2.718	34.86	95.39
10.8	10.8	2.718	35.86	100.00

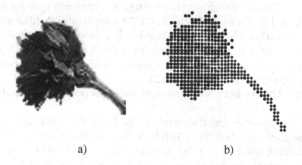

a)   b)

Fig. 2. Comparision image with standard (a) and proposed (b) screening technology (scale 10:1)

6   Conclusion

A new method of the forming a screening structure based on a periodic Ateb-functions is proposed. This structure is specially adapted for reproduction of fine protective graphical elements and halftones while printing, which improves the print quality greatly. The relationship between the Ateb-functions and the generalized superellipse is proved. Advantages of the method were shown in some experiment images. For improvement of this method we can construct asymmetric form of screening elements, and consider a screening point with an axis inclines at an angles $5^o - 15^o$. This method can be used for improving the effectiveness of protecting information on paper, plastic and other material media.

References

1. Kipphan, H.: Handbook of Print Media. Springer, Heidelberg (2001). ISBN: 3-540-67326-1
2. Bennett P., Romano, F., Levenson H.R.: The Handbook for Digital Printing and Variable - Data Printing, pp. 113–126. PIA/GATF Press, Pitsburgh, NPES (2007). ISBN: 978-5-98951-020-7
3. Sardjeva, R.: Investigation on halftoning methods in digital printing technology. Int. J. Graph. Multimed. (IJGM) 4(2), 1–10 (2013). ISSN: 0976 6448 (Print), ISSN: 0976 6456 (Online)
4. Sardjeva, R., Mollov, T.: Stochastic screening for improving printing quality in sheet fed offset. Int. J. Inf. Technol. Secur. 1, 63–74 (2012). ISSN: 1313-8251
5. Dronjuk, I., Nazarkevych, M., Medykovski, N., Gorodetska, O.: The method of information security based on micrographics. In: Kwiecień, A., Gaj, P., Stera, P. (eds.) CN 2012. Communications in Computer and Information Science, vol. 291, pp. 207–215. Springer, Heidelberg (2012)
6. Rosenberg, R.: The Ateb(h) functions and their proporties. Q. Appl. Math. 21(1), 37–47 (1963)
7. Sokolov, D.: Lame curve. In: Hazewinkel, M. (ed.) Encyclopedia of Mathematics. Springer, The Netherlands (2001). ISBN: 978-1-55608-010-4
8. ISO 14298: Management of security printing processes (2013).http://www.iso.org/iso/home/store/catalogue_tc/catalogue_detail.htm?csnumber=54594
9. Kuznetsov, Y.V., Zheludev, D.E.: Method of objective evaluation the fine detail distortion in process of screening. In: IARIGAI, N 35, pp. 347–353 (2008)

Author Index

Printed in the United States
By Bookmasters